Chicken Diseases Help
A Guidebook on
Chicken in Sickness and
Health

Norman Nelson

PUBLISHED BY:
Norman Nelson
Copyright © 2012

TABLE OF CONTENTS

Introduction

Farming chickens for fresh eggs and meat has been a way of life for people all over the world for thousands of years. As economic problems have increased over time, more of us are seeking ways to supplement our food provisions with healthy and affordable alternatives, like raising small flocks of chickens in our backyards.

Chicken Diseases Help is a comprehensive book for maintaining good health in your chickens, providing readers with symptoms, causes, preventative measures, and treatments for the most common diseases & sicknesses that can negatively affect the health of your flock, their offspring, egg production, and egg and meat quality.

This guidebook covers thirty of the most common infections and diseases occurring in chickens caused by: bacteria, fungi, parasites, and viruses, and provides readers with simple steps for keeping their chickens and homes healthy.

Chapter 1 - Causes of Chicken Sickness & Disease

Chickens are susceptible to sickness and disease like any other animal. The most common health problems are caused by: bacteria, fungi, parasites, and viruses. Diseases, infections, and illness are transmitted to chickens by parasites, like lice, mites, fleas, ticks, and worms, and through ingestion of living organisms, such as bacteria, bacterial toxins and fungi growing in their feed, water, and hay, and living in the soil where they forage. Chickens are also susceptible to develop respiratory illness from exposure to extreme cold, and the damp, toxic, and humid conditions that commonly occur within their housing as a result of the high quantity of moisture in their excrement.

Chickens are social creatures; they live, eat, drink, sleep, and share absolutely everything, including brooding and raising their young together. The only downside to this lifestyle is an outbreak of sickness in one chicken can result in the devastation of an entire flock. Understanding the different causes of disease and sickness in their natural environment, and how symptoms present, preventative measures, and treatment is critical to maintaining good health in a backyard flock or small farm, and the food they supply you and your family.

Bacteria

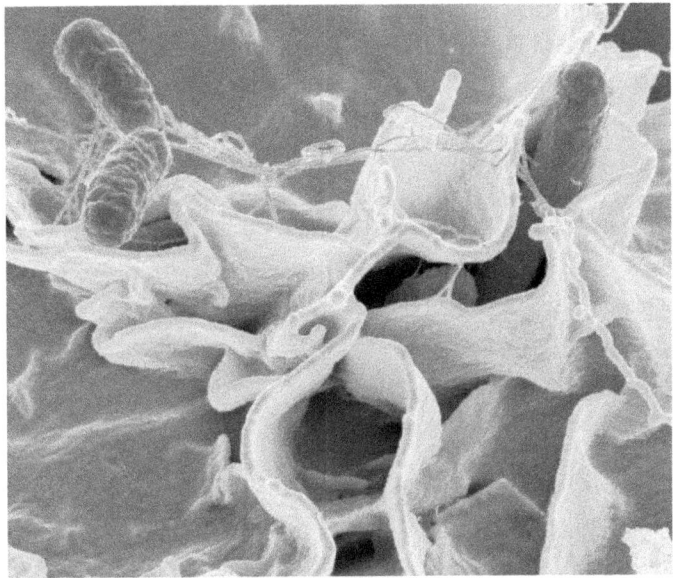

(1)

Bacteria are an enormous classification of microorganisms, separate from plants and animals. Under a microscope, these tiny organisms are prominently recognized for their distinctive shapes, which include spheres, spirals, and rod-like structures. There are more species of bacteria than plants and animals combined.

Bacteria are natural organisms living in the soil and water, on the surface and core of almost every habitat on the planet, and on and within the live body of all organic matter like in humans, animals, and plants. Bacteria are essential to life, providing powerful aid in human digestion, decomposition of organic matter, nutrient cycling and exchange, antibiotic production, fermentation of cheese, and sewage treatment. However, some bacteria are pathogenic, creating infections in humans and animals and even death.

Chickens are susceptible to infections and diseases from pathogenic bacteria growing in their natural habitat, food, and water sources. The bacterium is responsible for a large majority of health problems in chickens. Bacterial infections and diseases are most often spread from one chicken to another through the ingestion of infected feces.

Some of the most common bacteria-caused health problems include:

Bumblefoot, Ulcerative Pododermatitis
Campylobacteriosis
Erysipelas
Fowl Cholera
Fowl Typhoid
Infectious Coryza
Mycoplasmas
Necrotic Enteritis
Psittacosis
Pullorum, Salmonella
Ulcerative Enteritis
Yolk Sac Infection, Omphallitis

Bacterial Toxins

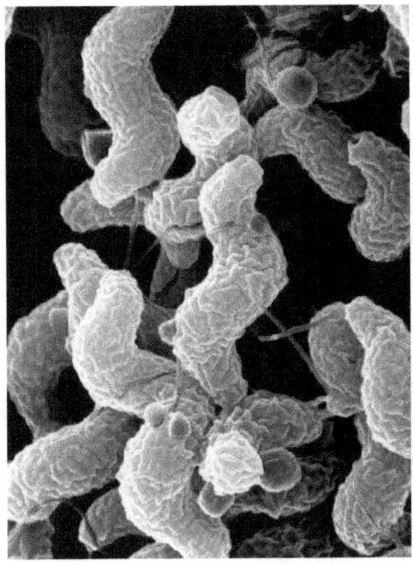

(2)

A toxin is a non-man-made poisonous substance produced by or within a living organism. Toxins may vary in strength from the temporary discomfort caused by the toxin in a bee sting, to deadly toxins contained in spider and snake bites that attack red blood cells or the nervous system causing death immediately upon contact.

Small organisms like bacteria, fungi, and viruses also produce toxins, some highly dangerous and deadly, and commonly attributed to food or blood poisoning.

One of the most dangerous bacterial toxins that cause food poisoning, extreme pain, illness and even death in humans is equally harmful to chickens:

Botulism

Fungi

(3)

A fungus is classified as a living organism, separate from plants, animals, and bacteria. Through mycology (the study of fungi), an estimated 1.5 million fungi species exist with a life cycle more closely resembling animals than plants.

Fungi are natural organisms found in the soil and involved in the decomposition of living matter, such as animals, plants, and other fungi. They often go undetected due to their small size until they start fruiting into visible molds. Fungi play a vital role in the decomposition of organic matter, as well as in nutrient cycling and exchange. While certain fungi are utilized in the production of antibiotic medicine and other consumer products, many others are toxic and pathogenic, leading to sickness, disease, and even death in humans and other animals.

The most common chicken health problems caused by fungi or fungus include:

Aspergillosis
Moniliasis, Yeast Infection, Thrush

Parasites

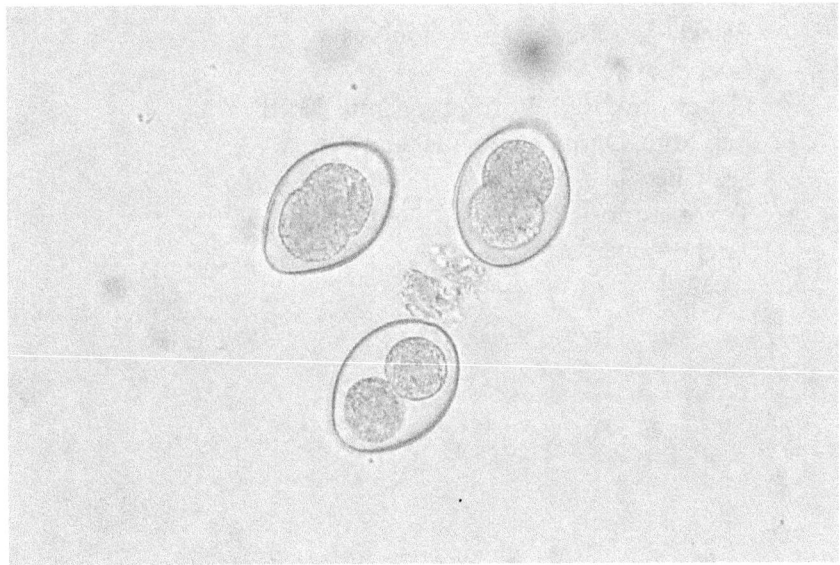

(4)

Parasitism is a non-mutual relationship between two living species, where the parasitic organism grows, feeds, and finds shelter on or inside a different organism known as the host. The parasite benefits at the expense of the host, often depleting essential nutrients needed to maintain sound health. Parasites infect their hosts by burrowing into and biting the skin or through ingestion, and they can easily transfer from one host to another.

The most commonly known parasites include lice, mites, fleas, ticks, and worms. These parasites cause irritation, deplete the host of essentials like food, water, heat, and blood necessary for host survival, and spread dangerous pathogens, leading to discomfort, disease, and even death in humans and animals.

Intrusive parasites that invade the internal organs of chickens cause a wide range of health problems, which include:

Blackhead Disease, Histomoniasis
Coccidiosis
Gapeworm, Red Worm, Syngamus Trachea
Red Mite, Dermanyssys Gallinae
Scaly Leg
Toxoplasmosis
Trichomoniasis

Virus

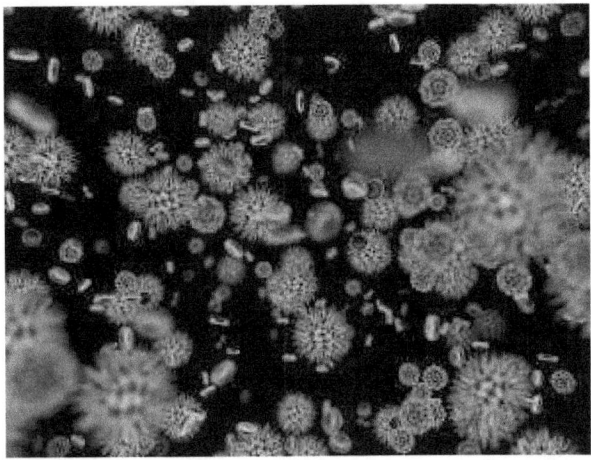

(5)

A virus is a small infectious agent that only replicates itself inside the living cells of an organism. A virus may also be a pathogen and can replicate and cause damage, even death to its host. A virus can spread quickly through a flock. Viral infections are automatically attacked by the immune system of the host, which often eliminates the infecting virus. While antibiotics are of no use against viruses, many vaccines are available for some prevention and treatment.

Viral infections and diseases adapted and mutated from different species of birds, animals, and even humans, dangerous to the health of your chicken flock and other livestock include:

Bird Flu, Avian Influenza
Fowl Pox
Gallid or Avian herpesvirus 1, GaHV-1, Infectious Laryngotracheitis, LT
Infectious Bronchitis

Infectious Bursal Disease, Gumboro
Lymphoid Leukosis
Marek's Disease
Newcastle Disease

Chapter 2 - Chicken Disease Summaries

In this chapter, readers will discover comprehensive and detailed summaries of thirty of the most common chicken diseases, illnesses, and infections that can adversely affect the health and production levels of chickens in both small backyard farms and commercial poultry houses. All the "diseases" covered in this book are not region-specific but rather prevalent among chickens worldwide.

Each "disease" summary provides the following:

- Type of Health Problem:
  Illness
  Infection
  Disease

- Cause of the Illness, Infection, or Disease:
  Bacteria
  Bacterial Toxin
  Fungi
  Parasite
  Virus

- Source of the Illness, Infection, or Disease:
  Environmental conditions for growth and transmission
  Transmission factors between chickens in a flock
- Symptoms of the Illness, Infection, or Disease:

Early indicators of the onset of infection
Behavioral indicators of the illness, infection or
disease
Physical symptoms

- Prevention of the Illness, Infection, or Disease:
Medicines available for prevention
Medicines available for treatment
Methods for prevention in environment
Methods to prevent spread of illness, infection, or
disease to chickens within a flock.

References for every illness, infection and disease have been
supplied at the end of the book with its corresponding
number in order to provide readers with access to
photographs and additional information.

Aspergillosis

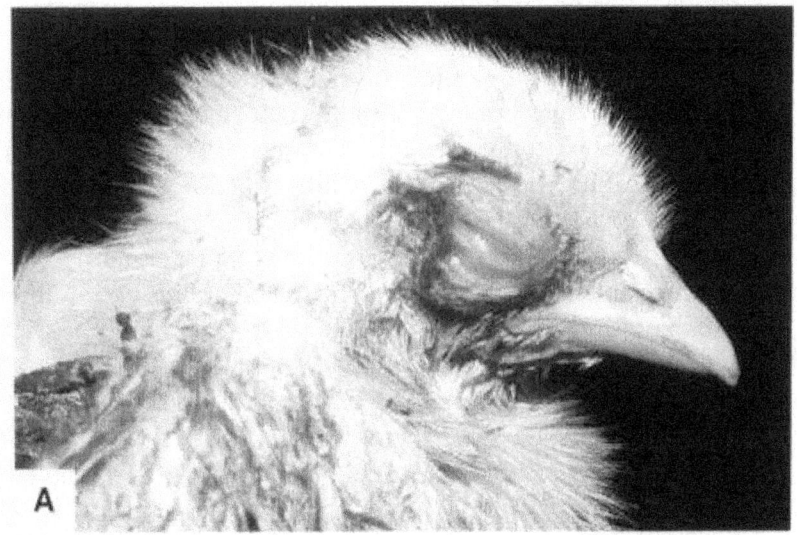

(Panophthalmitis -Asperagellosis)   OldVeT.com

(6)

Type: Dangerous infection in chickens, most commonly attacking the lungs.
Cause: Fungi, of Aspergillus.
Source: Consumption of contaminated food sources, moldy grain and feed.
Symptoms: Repeated coughing, coughing up blood, fever and chills, chest pain, and difficulty breathing. Aspergillosis may also be fatal, causing flock losses.
Prevention: Maintaining proper feed storage, free of moisture and mold growth is critical to prevent exposure and contamination.

Aspergillosis is a common and potentially dangerous infection in many birds, including chickens, attacking the

19

lungs, tissue and organs. Fungi, like aspergillus, are naturally found in soil, decaying vegetation, hay and grain, and grow rapidly in damp environments commonly linked to improper storage of feed. While mold is a natural microorganism used in the production of some foods and medicines, the aspergillus fungus is a toxic mold. When chickens ingest this deadly mold it often forms into a fungus ball within their lungs, creating severe respiratory inflammation, before spreading to other tissues and organs, potentially leading to fatality. The most common indicators of an aspergillosis infection in chickens include: difficulty breathing, continuous coughing, and/or coughing up blood.

Aspergillosis has been connected with large flock fatalities, caused by feeding chickens moldy grain, and even leaving piles of moldy waste grain, contaminated with aspergillus fungi, unattended.

The most effective way to keep a flock safe from contracting the aspergillosis infection is to buy, store and serve feed properly. Keeping feed dry and protected from moisture keeps this fungi from growing into a dangerous and potentially deadly mold.

"Bird Flu" or Avian Influenza, HPAI - highly pathogenic avian influenza

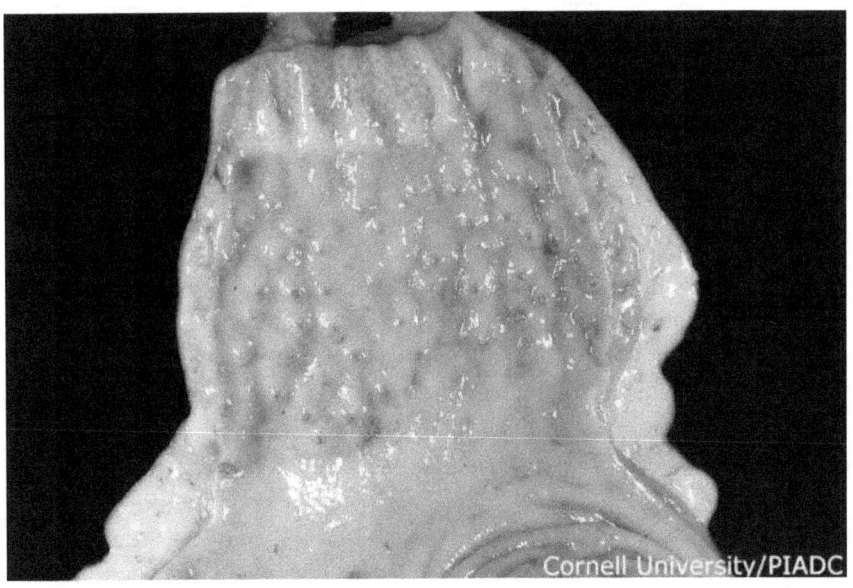

(7)

Type: Dangerous, potentially pandemic influenza or flu that has adapted to birds, as well as humans, and one of the top concerns for world-wide disease control.

Cause: Virus, HPAI A (H5N1).

Source: Spread by birds and other mammals rapidly through fluids: saliva, nasal secretions, feces and blood.

21

Symptoms: During infectious periods, chickens will not exhibit any flu-like symptoms. Often symptoms only arise after the flu virus, usually adapted for another bird species, mutates into a form that wipes out the entire flock within days, and continues to spread to other flocks, killing them, and so on. Mortality rates of chickens or other birds infected with the H5N1 are 100%.

Prevention: Vaccines may be used for prevention for many strains, including some of the avian influenza H5N1 varieties.

Avian influenza," commonly known as "bird flu," is a severe illness caused by various strains of influenza or flu-type viruses that have adapted to chickens, other bird species, as well as many mammals and humans. All influenza viruses that birds can contract fall under the classification of "influenza A virus."

Infected chickens, acting as carriers of this deadly virus, may not exhibit flu-like symptoms but can quickly spread the virus to the entire flock through their bodily fluids. Due to its often undetectable nature and rapid mutation, the virus can wipe out an entire flock and spread to others within days, posing a significant risk to both poultry and potentially the human population. The highly pathogenic H5N1 strain, first identified in Asia in 2003, has since surfaced on almost every continent and in many countries. Disease control centers worldwide closely monitor this virus due to its devastating potential for chickens, birds, mammals, and humans alike.

While some vaccines are available for certain influenza A virus strains, there is currently no known cure. When acquiring new chickens or a flock, it is crucial to ensure they come from a reputable and legal poultry dealer. Illegal poultry traders are unregulated and often driven by profit rather than safety.

Black Head or Histomoniasis

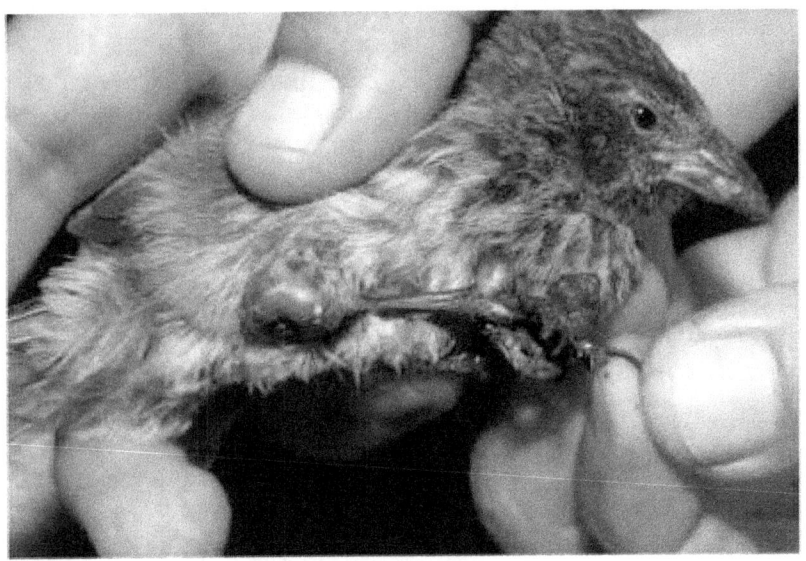

(8)

Type: Dangerous intestinal and liver disease in poultry birds, caused by parasitic roundworm infection.
Cause: Parasite, Histomonas meleagridis.
Source: Consumption of soil and earthworms contaminated with parasitic eggs, usually transmitted through feces.
Symptoms: Depression, loss of appetite, increased thirst, reduced egg production, yellow watery diarrhea, and may cause discoloration of the head, often bluish or black in color.

Histomoniasis is a serious disease and has a high fatality rate in poultry birds.

Prevention: There are currently no preventative or treatment drugs available for this disease. Maintaining sanitary indoor and outdoor conditions with frequent litter changes is the most proactive approach for prevention.

This tiny parasite produces tiny roundworm eggs in the soil that can remain dormant for four years until ingested. When these tiny parasitic eggs are ingested by chickens, they hatch and attach to the chicken's intestines, before infecting the liver, creating internal lesions. Transmission of this parasite to chickens can also occur through the ingestion of earthworms that consumed the tiny parasitic eggs. These parasites are commonly present, and easily transmittable from one chicken to another, through contaminated feces.

It is sometimes called the blackhead disease because of the skin discoloration that may occurs on the head of some of the infected chicken, appearing blue or black.

Chickens infected with histomoniasis may also appear tired and depressed, experiencing a loss of appetite and increased thirst. Egg production in infected chickens can decrease dramatically or stop altogether. Feces turn yellow and watery and may contain visible parasite eggs, and even tapeworms.

Sick birds should always be separated from healthy birds to prevent the spread of this dangerous parasite throughout a flock. There have been many trial drugs for the treatment of parasitic worms; however, no actual preventative or cure exists for this disease. Maintaining a clean environment is the best defense against this disease.

Botulism

(9)

Type: Rare but serious paralytic illness caused by ingesting a bacteria toxin.

Cause: Bacteria Toxin, Clostridium botulinum, or Botulin Toxin.

Source: Consumption of contaminated food.

Symptoms: Paralysis or impaired muscle functions in the legs, wings, neck and eyelids. Botulism can lead to suffocation if respiratory and breathing muscles lose their ability to contract normally.

Prevention: The best preventative measure is the proper disposal of tainted food, particularly foods that may have been improperly canned or have a broken can seal. These

25

foods should never be added to compost or fed to animals, as they can also make them sick.

Botulism is a severe illness that poses a danger to humans, mammals, and birds, including chickens. The botulin toxin is one of the most potent and powerful known toxins, causing paralysis and potential death upon ingestion. Once swallowed, botulism attacks the intestines and impairs muscle functions. Chickens will exhibit signs of weakness and reduced muscle tone in their legs, wings, neck, and eyelids. Muscles become limp, hindering movement, including respiratory and breathing functions. If the respiratory muscles lose their ability to contract normally, suffocation can occur.

Botulism cannot be spread from chicken to chicken through contact, but it is hazardous because if contaminated feed is ingested, it can affect an entire flock. To prevent botulism, it is essential to kill the toxin by boiling food and cooking foods at high temperatures. Any food waste, especially home-canned goods and foods stored in metal cans that do not appear properly sealed, and may even be bulging due to gas, should be disposed of correctly to ensure chickens do not ingest this deadly toxin.

"Bumblefoot" or Ulcerative Pododermatitis

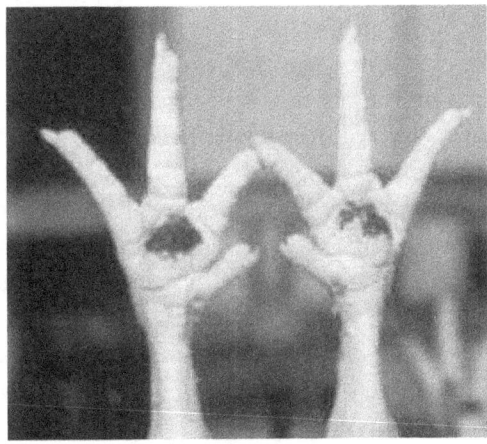

(10)

Type: Bacterial infection occurring through open wounds in the feet. Inflammation causes swelling, pus, and increased size of lesions. Loss of foot, and even death possible, if not properly treated.
Cause: Bacteria, like Staphylococcus aureus, Escherichia coli
Source: Open foot wounds from walking on rough, sharp surfaces.
Symptoms: Bumblefoot may initially present as a small reddened areas on the foot, a wound, and even in foot abnormalities, resulting from untreated infected wounds.

Prevention: Once an open wound or a small reddened area indicating inflammation is discovered, topical antiseptics and/or oral or intravenous antibiotics should be used as proper treatment. Severe infections can lead to loss of foot and death if left untreated.

Bumblefoot is a bacterial infection that causes inflammation in the feet of birds. This inflammation is common to many domestic birds, especially chickens that spend most of their time inside wire cages. A chicken's natural habitat is soft ground, soil, and grass. When they are forced to continuously walk on sharp objects and hard metallic surfaces, chickens will often develop small wounds on the bottom of their feet. These wounds are highly prone to infection since chickens also commonly walk on top of their feces.

Bacteria like Staphylococcus aureus and Escherichia coli when they enter the wound causes bumblefoot or ulcerative pododermatitis. Bumblefoot may initially present as a small reddened area on the foot, indicating inflammation. Once a foot wound is visible, it often becomes severely infected fast. Infected wounds may require lancing and draining to remove pus, and must be treated with an antiseptic and wrapped for several weeks, to keep the wound sterile while it heals. For severely infected wounds, oral or intravenous antibiotics are typically required to fight off this bacterial infection. Bumblefoot infections can lead to severe distorting of the foot structure and toes, and complete loss of foot function, and even death.

By maintaining good chicken runs with access to walk in their natural habitat in outdoor runs, bumblefoot may be prevented altogether. If wounds present, immediate treatment, sterile dressings for several weeks, and clean litter in indoor runs can keep chickens and their feet healthy.

28

Campylobacteriosis

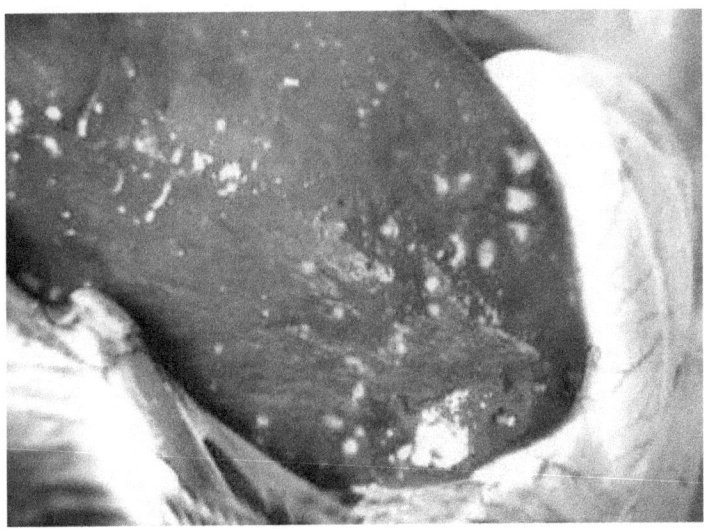

(11)

Type: Infection caused by food-borne bacteria.
Cause: Bacteria, Campylobacter
Source: Consumption or contact with contaminated food, or infected feces.
Symptoms: Diarrhea, watery or bloody feces are common indicators, and may be accompanied by fever, severe dehydration and abdominal pain.
Prevention: The best preventative measure is the proper disposal of tainted animal meat and by-products. These foods

should never be added to compost or fed to animals, as they can also become sick.

Campylobacteriosis is a serious infection that can be dangerous to humans, mammals, and birds, including chickens. The food-borne bacteria, Campylobacter, is the most common bacterial infection in humans and is most often transmitted through contact with animals farmed for their meat, poor handling, and undercooking poultry. This bacterial infection causes diarrhea, watery or bloody feces, fever, severe dehydration, and abdominal pain, in chickens and humans alike. Chickens may also develop a gut injury to the infected tissue in their intestines. As a result of severe infections, chickens can also experience paralysis in their legs, and loss of respiratory and breathing functions, and even death.

Campylobacteriosis bacterium is transmitted by fecal-oral, ingestion of contaminated food, often undercooked or improperly handled poultry, and through contaminated drinking water. This means rapid transmission from one chicken to an entire flock.

Campylobacteriosis bacterium can be killed or reduced by boiling food, and cooking foods at high temperatures. Any food waste, especially old or spoiled animal meat should be properly disposed of so chickens do not ingest this bacterium.

This cholera-like intestinal infection usually runs its course within two weeks, and in many cases, can be treated by hydration alone. Past antibiotic treatments have been banned as ineffectual. While Campylobacteriosis can cause a major outbreak of sickness in a flock, it is rare for this infection to cause death.

Coccidiosis

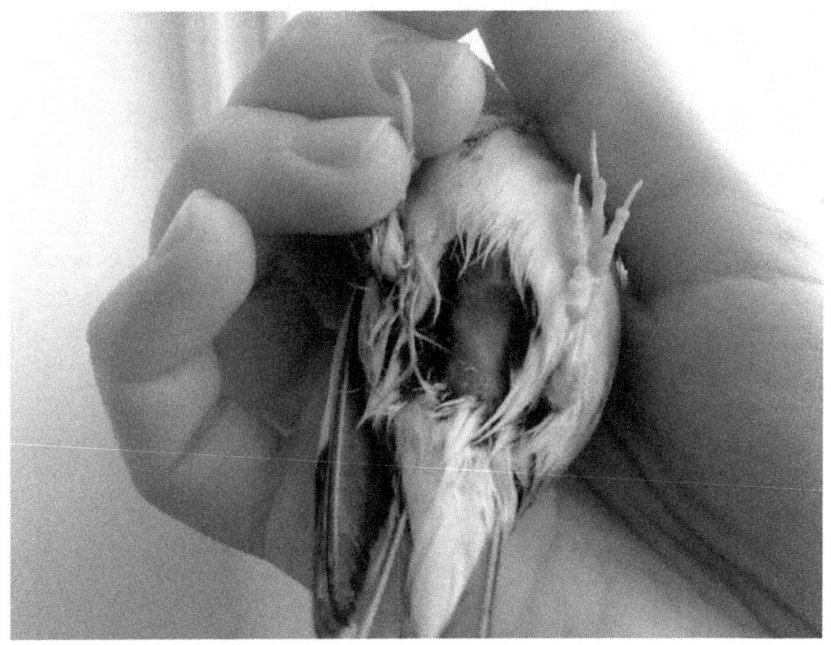

(12)

Type: Intestinal tract disease caused by parasitic coccidian infection.
Cause: Parasite, Coccidian Protozoa
Source: Consumption or contact with contaminated feces.
Symptoms: Diarrhea or watery feces is the primary indicator. Young or older chickens can suffer more severe symptoms, including death.

Prevention: Zoalene, Zoamix, Coccidine A, or Coccidot is a feed additive for poultry, used to prevent and treat coccidiosis infections.

Coccidiosis is a common disease of the intestinal tract caused by parasitic infection, affecting chickens. The protozoa parasite, Coccidian, is microscopic, spore-forming, and must live and reproduce within an animal cell. These tiny parasites, once ingested, make their way to the intestinal tract where they reproduce, form small cysts, and are released in the chicken's feces.

Coccidiosis spreads from chicken to chicken through contact and ingestion of contaminated feces, and can quickly render an entire flock of chickens sick in a short amount of time.

While chickens with mild infections of this parasite do not show any symptoms, diarrhea is the primary indicator for coccidiosis infections. Bloody diarrhea may present in severe cases.

Zoalene, Zoamix, Coccidine A, or Coccidot is a chicken feed additive, specifically formulated for poultry, used to prevent and treat coccidiosis infections. This additive is highly effective against seven main strains of this parasite.

Erysipelas, "Holy Fire"

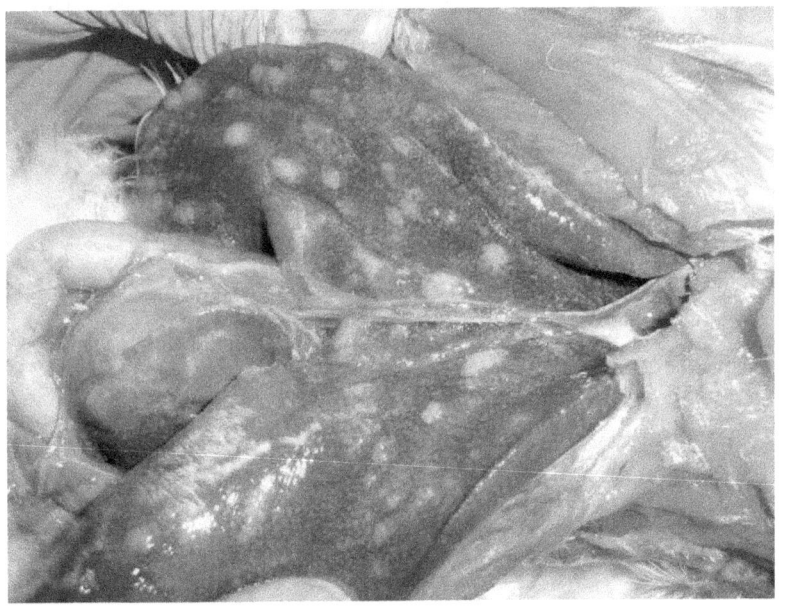

(13)

Type: Bacterial infection of the skin; red rash.
Cause: Bacteria, Erysipelothrix rhusiopathiae
Source: Open wound, scratch, or minor skin trauma.
Symptoms: Skin inflammation, redness, rash and swelling, wound enlargement and decay, fever, chills, shaking, and vomiting.
Prevention: Penicillin is the preferred treatment.

Erysipelas is a bacterial infection of the skin and lymphatics caused by the bacteria, Erysipelothrix rhusiopathiae. This bacterium attacks the skin but also travels throughout the body through white blood cells, and can cause a chicken to become seriously ill in a short amount of time. This bacterial infection often presents in open wounds, even a small scratch on the chicken's body, and can cause severe fever, chills, shaking, and vomiting within a few days after infection. Because this bacterium travels through the blood, it can cause severe sickness in chickens and humans alike, and even death if left untreated.

Wounds exposed to this bacterium may become highly infectious to other chickens, and even develop gangrene where the infected tissue dies and begins to decay.

Penicillin is the preferred treatment for this bacterial infection and can clear up illness symptoms within a few days. The skin requires additional time to heal and may take several weeks to return to normal.

All scratches and open wounds should always be addressed at the moment they are noticed, and can often be healed with topical antiseptic and clean dressings before infections like erysipelas can set in.

Fowl Cholera

(14)

Type: Infectious and deadly bacterial disease.
Cause: Bacteria, Pasteurella multocida
Source: Ingestion of contaminated food or water.
Symptoms: Convulsions, respiratory and breathing problems, fever, and sudden death.
Prevention: Healthy and sick birds should be separated as quickly as possible. Chlortetracycline, oxytetracycline and sulfaquinoxaline may be added to feed or water for treatment of the remaining flock.

Fowl cholera is an infectious disease caused by bacteria in birds, including chickens, and has been associated with migrating birds.

Transmission of this disease is believed to occur through ingestion of the Pasteurella multocida organism, either in contaminated food or water supply. An outbreak can occur suddenly, and is often undetected until several chickens fall ill and die quickly. This infectious disease can spread quickly from sick chickens to healthy ones.

Chickens infected with fowl cholera may exhibit symptoms including: convulsions, fever, shaking and shivering, sneezing, runny nasal fluid and yellow-colored diarrhea, swollen bellies, weight loss and listlessness.

Fowl cholera outbreaks often result in losses, and separating sick birds from healthy one is the top priority. Chlortetracycline, oxytetracycline and sulfaquinoxaline may be added to feed or water for treatment of the remaining flock. Chicken carcasses should be burned to prevent additional outbreaks in other birds and animal species.

Fowl Pox

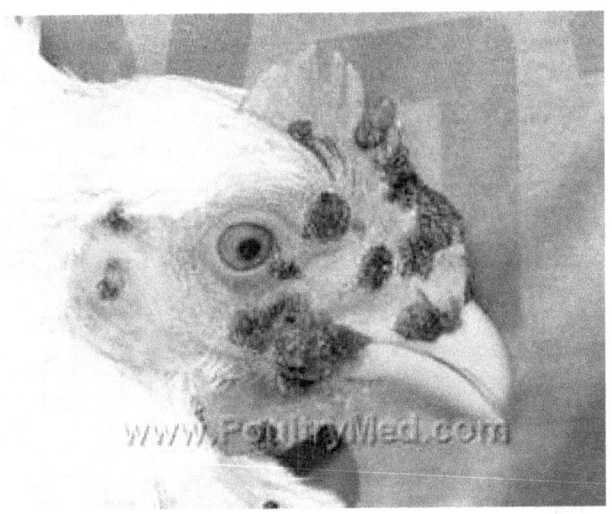

(15)

Type: Viral disease affecting birds throughout the world.
Cause: Virus, Avipoxvirus
Source: Spread by mosquitoes and other biting insects, through open wound contamination, inhalation and ingestion.
Symptoms: Fowl Pox causes pustules, sores and lesions on the head; to the comb, wattle, beak, mouth, throat and respiratory system, and difficulty eating and breathing.
Prevention: Vaccine is used to prevent fowl pox.

Fowl Pox is an eruptive avian viral disease caused by the virus Avipoxvirus, which affects poultry birds worldwide. The disease leads to the development of pustules, sores, and lesions on the bird's body, commonly found on the head, comb, wattle, beak, mouth, throat, and respiratory system. In some cases, lesions can also appear on areas of the body that are free of feathers, including the legs.

Fowl Pox is typically contracted through biting insects like mosquitoes, which transmit the virus to birds. It can also occur through open wound contamination. Many chickens may develop small sores and eventually recover completely within a few weeks. However, in more severe cases, the disease can be transferred from sick birds to healthy ones through inhalation and ingestion of the virus, leading to internal infection in the mouth, throat, and respiratory system. The survival rate of chickens with severe internal pustules and lung infection is lower than for those exhibiting only external symptoms.

To prevent Fowl Pox, chickens are vaccinated with the Fowl Pox vaccine. Sometimes, a vaccine based on pigeon pox virus is used for prevention as it provides cross-protection against Fowl Pox.

Fowl Typhoid

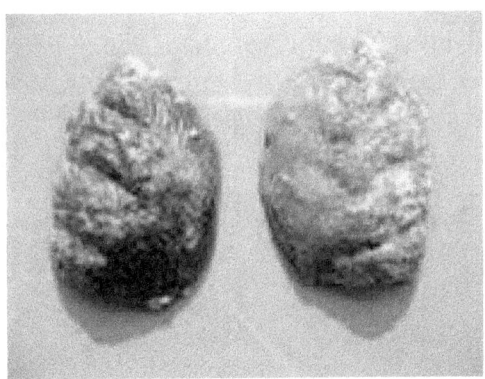

(16)

Type: Chronic bacterial disease affecting mature birds throughout the world.
Cause: Bacteria, Salmonella gallinarum.
Source: Consumption of contaminated feed and water, and through egg transmission.
Symptoms: Fever, pale combs and wattles, yellow diarrhea, and listlessness are the common indicators. Respiratory distress and death may also occur.
Prevention: Neomycin or sulfaquinoxaline may be added to feed for prevention

Fowl Typhoid is a chronic bacterial disease that affects mature poultry birds throughout the world. This disease is caused by the bacteria, Salmonella gallinarum, causing adult birds fatigue and listlessness, fever, loss of appetite, yellow diarrhea, and dehydration. Combs and wattles may lose their color, appearing pale.

Fowl Typhoid is now rare in commercial poultry, but can still affect the chickens of small farmers. This bacterial disease affects adult or mature chickens almost exclusively. Brown egg layer breeds are more susceptible to contracting this disease than white egg producers.

Fowl Typhoid is spread through the consumption of contaminated feed or water and infected eggs. Mortality rate can be as low as 10%, and as high as 100% for infected chickens.

Neomycin or sulfaquinoxaline may be added to feed for prevention.

Gallid or Avian Herpesvirus 1, GaHV-1"Avain Infectious Larynogotracheitis, or LT"

(17)

Type: Highly contagious infectious disease of the larynogotracheitis, or LT.

Cause: Virus, Herpesvirdae, GaHV-1

Source: Contact with infected fluids: salvia, nasal secretions and feces.

Symptoms: Coughing, sneezing, discharge from the eyes and nose, respiratory and breathing distress, listlessness and fatigue, and can affect egg production, resulting in abnormal eggs, or eggs with thin shells.

Prevention: Isolation and intensive cleaning are the best preventative measures.

Gallid or avian herpesvirus 1 is a highly infectious disease causing inflammation of the larynx and trachea in birds caused by the virus, Herpesviridae, or GaHV-1. Avian herpesvirus 1 or LT is a highly contagious disease among bird flocks, including chickens. While death is not common, an outbreak typically results in a quarantine of the farm to prevent spreading of virus throughout the flock and to other bird species.

GaHV-1 presents in chickens through coughing and sneezing, discharge from the eyes and nose, respiratory and breathing problems, listlessness, and fatigue. GaHV-1 causes severe inflammation of the larynx and trachea and can result in airway obstruction due to internal swelling of the throat. This viral infection is transmitted through fluids and can spread quickly to an entire flock within two to eight weeks.

Isolation of sick birds, disinfecting and cleaning the chicken housing, equipment, water supply, and common areas is also critical to contain this virus until it runs its course. While death can occur, the mortality rate of GaHV-1 is relatively low.

Gapeworm, Red Worms, or Syngamus Trachea

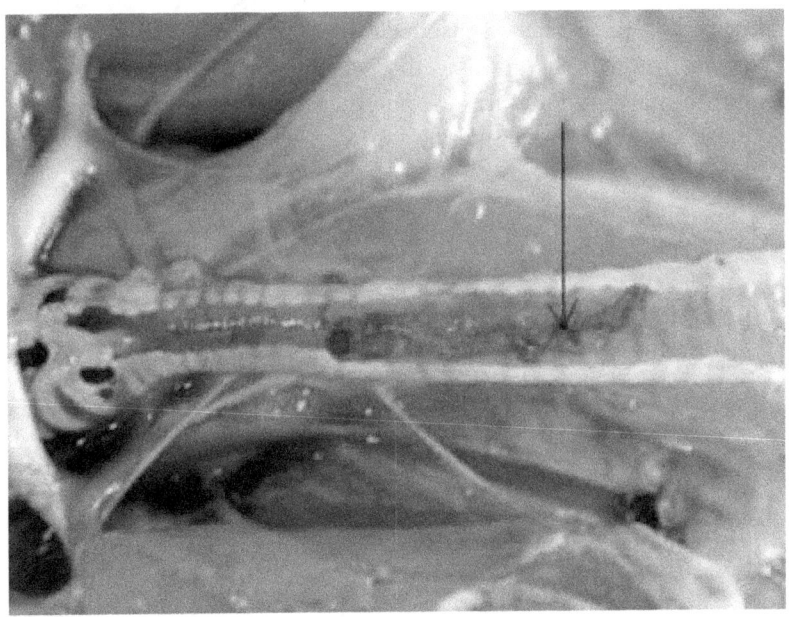

(18)

Type: Disease of the trachea in domesticated birds, caused by parasitic red worms.
Cause: Parasite, Gapeworm or Red Worm
Source: Consumption of contaminated soil or earthworms.

Symptoms: Trachea and throat swelling, lung inflammation, coughing up mucus and blood, pneumonia, obstructed airway and even suffocation.

Prevention: Drugs, Ivermectin and Fenbendazole are used with success to treat gapeworm infections. Disinfecting, and maintaining clean litter, housing and feed equipment is highly effective as a preventative to stop the spread of this parasite.

Gapeworm is a disease of the trachea and throat in young birds, common to chickens, caused by ingesting parasitic gape or red worms. This disease is common in young chickens and often presents through swelling of the throat, lung inflammation, coughing up mucus and blood, pneumonia, airway obstruction, and even suffocation.

The parasitic gapeworm attaches itself to the trachea of an infected bird and lays eggs that are swallowed and passed through feces. Other chickens ingest the gapeworm eggs and become infected. Earthworms may also consume the gapeworm eggs and transmit this disease to chickens that may consume them. The gapeworm stays attached to the trachea blocking the airflow with mucus and swelling, causing chickens to gasp for air. Symptoms typically present within a week or two of infection. Chickens often stop eating and drinking and repeatedly shake their heads to clear the object from their airway.

The drugs, Ivermectin and Fenbendazole, are effective in treating gapeworms in chickens. Disinfecting chicken housing, feed and water equipment, litter, and soil is also highly effective in containing and eliminating this parasite.

Infectious Bronchitis, IB

(19)

Type: Contagious viral upper respiratory disease of chickens, affecting birds throughout the world.

Cause: Virus, Coronavirus.

Source: Consumption of contact with fluids of infected birds: saliva, nasal and eye secretions, and feces.

Symptoms: Coughing, sneezing, nasal secretions, respiratory and breathing distress.

Prevention: Many vaccines exist for chickens. Disinfecting and maintaining clean litter, housing and equipment can eliminate the virus from spreading.

Infectious bronchitis or IB is a viral upper respiratory disease in chickens. IB is caused by a coronavirus that infects the upper respiratory system in animals and birds.

IB initially presents through repeated coughing and sneezing, nasal and eye secretions, respiratory or breathing distress, and even death. The virus is transmitted from chicken to chicken through their fluids during coughing and sneezing. Chickens will present with symptoms within one or two days of infection. This respiratory disease typically runs its course in a week. Mortality for chickens with IB is relatively low unless airway blockage occurs or the chicken is weakened by other ailments.

Many different vaccines exist for different strains of IB, according to chicken type, broiler, layer, etc. Disinfectants and sunlight are highly effective in killing this virus to prevent its spread.

Infectious Bursal Disease, IBD, or Gumboro

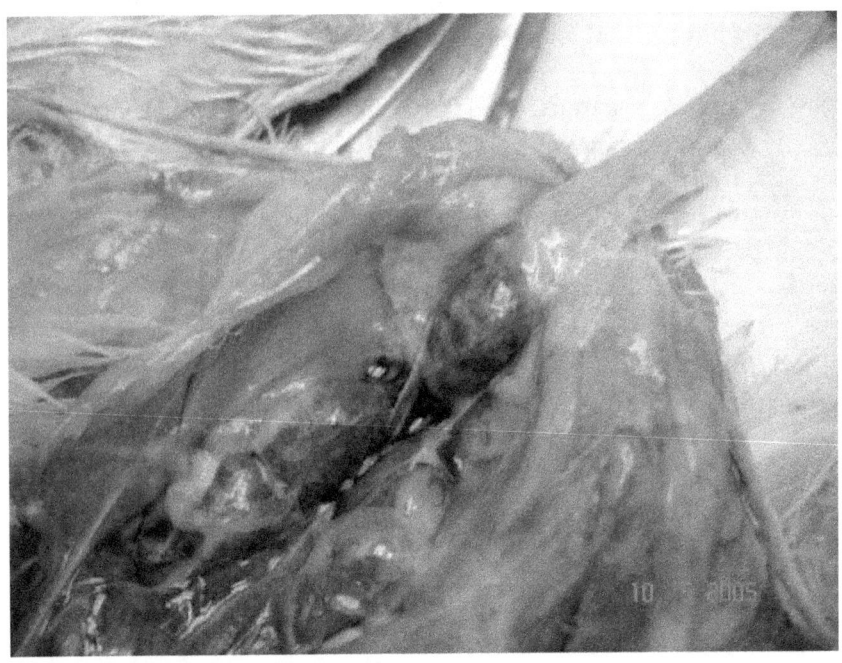

(20)

Type: Infectious viral disease affecting the bursal of fabricius organ in young chickens, worldwide.
Cause: Virus, IBDV (infectious bursal disease virus)
Source: Ingestion and contact with feces of infected chicks.

Symptoms: Watery and bloody diarrhea, and a swollen and blood stained vent.
Prevention: Vaccinations are not effective during outbreak.

Infectious Bursal Disease, IBD, or Gumboro Disease, is a viral disease affecting young chickens worldwide. The IBDV virus attacks the bursa of Fabricius, an internal organ found inside the cloacal opening used for intestinal, reproductive, and urinary functions in chickens. The bursa of Fabricius contains lymphoid cells and is primarily an internal tissue for absorption and protection, critical for the development of healthy blood components and immune systems.

This virus is transferred between chicks through the ingestion of infected feces by other chicks. Chicks may present with symptoms including watery and bloody diarrhea and a swollen and blood-stained vent as early as two weeks of age. Because this virus attacks immune development, some infected chicks will die within three to six weeks of age; mortality rates can be 40% or higher. Infected birds can continue to excrete this virus for two weeks after infection. Chicks, eight weeks or older, have developed immune systems and are more resistant to infection.

Vaccination is not effective in an outbreak, and no known treatment is available. Disinfecting litter, housing, and water supply are critical to prevent further spread as many strains have developed over time.

Infectious Coryza, IC, Coryza, Roup

(21)

Type: Upper respiratory disease of the sinuses, trachea and lungs, affecting poultry birds worldwide.
Cause: Bacteria, Haemophilus paragallinarum
Source: Inhalation and ingestion of bacteria infected food and water.
Symptoms: Sneezing, coughing, nasal swelling and discharge, watery eyes, respiratory and breathing difficulty.
Prevention: Sulfonamide or antibiotic treatments are available. Sulfadimethoxine is the safest and most widely used prescription drug for Coryza. Disinfecting litter, housing and

feed and water equipment is highly effective for prevention. Care should also be exercised when introducing new birds into an existing flock.

Infectious Coryza is an infectious bacterial disease affecting upper respiratory system, sinuses, trachea, and lungs of birds, primarily chickens. Coryza or IC is caused by the bacteria, Haemophilus paragallinarum, causing chickens to experience swelling of sinuses and eyes, sneezing, coughing, nasal and eye discharge, often a thick mucus with a foul odor, trachea swelling, and lung inflammation.

Coryza can afflict young and mature chickens, typically 14 weeks or older. Coryza most often spreads when carriers are introduced to a healthy flock and can infect the entire flock rapidly, as this bacterium can survive in fluids expelled by coughing and sneezing and in shared water and feed sources. Care should always be exercised when introducing new birds into existing healthy flock.

Once infected, chickens present with symptoms within one to three days. While death can occur, especially for birds with compromised immune system, and may be sick with other ailments, mortality is typically less than 20%. This respiratory disease often runs its course in four to 12 weeks.

Sulfonamide or antibiotic treatments are available for Coryza outbreaks. Sulfadimethoxine is the safest and most widely used prescription drug to treat this disease. Water and feed additives can reduce symptoms and the spread of this disease. The most effective preventative and containment includes disinfecting litter, housing, and equipment and maintaining a sanitary environment free of bacteria.

Lymphoid Leukosis

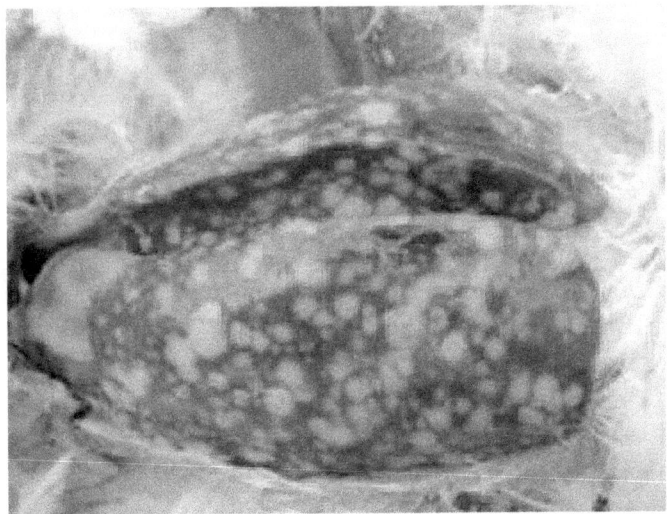

(22)

Type: Internal tumor causing virus in chickens and other birds.
Cause: Virus, Avian Sarcoma Leukosis Virus, ASLV
Source: Passed on in genetics through offspring.
Symptoms: External symptoms for lymphoid leucosis tumors include listlessness, fatigue, loss of appetite, and death.
Prevention: Acquire chickens from responsible farmers or dealers.

Lymphoid leucosis is a cancer occurring in chickens, and other bird species, caused by the avian sarcoma leucosis virus, or ASLV.

ASLV is a retrovirus that infects the cells that form connective tissues in chicken embryos. Retroviruses are old viruses passed on undetected to new generations through their offspring and have no known cure but have been effectively bred out of circulation over time.

Different forms of cancer evolve from ASLV. The primary diseases include:
●Lymphoblastic – cancer of the lymphoblast cells found in the bone marrow
●Osteopetrotic – cancer of the bones, where bones become hard

Once chickens are infected with this virus, they can present with a wide range of tumors in various organs including bone marrow, bones, liver, bursa of Fabricius, and abdominal organs. Chicken suffering from lymphoid leucosis often present with symptoms of listlessness, fatigue, loss of appetite, and movement, followed by death.

There has been extensive testing on the ASLV by scientists and health organizations to understand and combat this retrovirus. While carrier birds infected with this virus do not present with noticeable symptoms, obtaining birds from reputable dealers and health-conscious farmers is an effective preventative to avoid cancer development in existing flocks.

Marek's Disease, Marek's Disease Virus, MDV, Gallid herpesvirus 2, GaHV-2

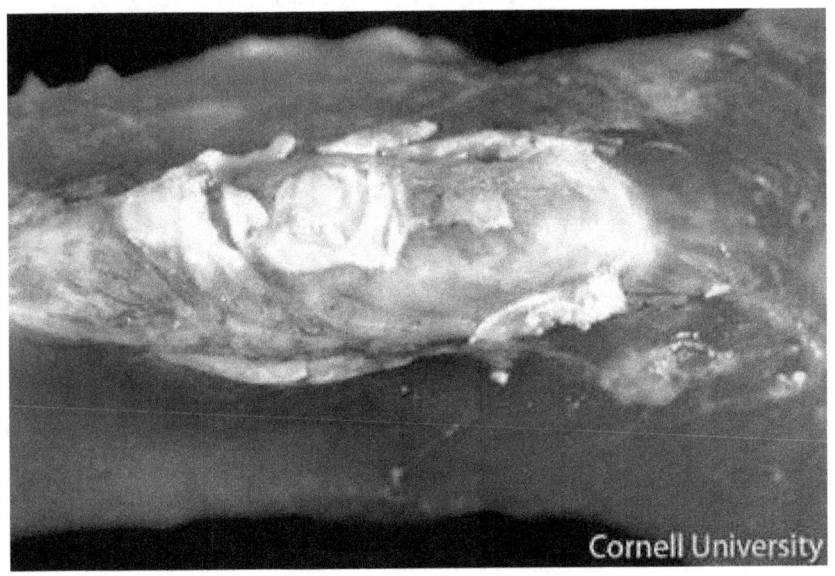

(23)

Type: Infectious tumor and lesion causing viral disease of the nerves and internal organs in chickens.
Cause: Virus, Gallid herpesvirus 2, GaHV-2
Source: GaHV-2 is spread in skin dander, through inhalation of infected particles.
Symptoms: Lesions and nodules, eye bulging, blindness, paralysis, reproductive failure, and even death.

53

Prevention: Vaccination can prevent the development of tumors in infected chicks, but does not eliminate the disease.

Gallid or avian herpesvirus 2 is a highly infectious viral disease in chickens caused by the virus, Gallid herpesvirus 2, or GaHV-2. This disease may also be referred to as Marek's Disease, and the virus as Marek's Disease Virus, or MDV.

Avian herpesvirus 2 infects healthy cells, nerves, and organs of chickens, causing eye bulging and abnormalities, internal tumors or nodules, lesions, paralysis, and even death.

The GaHV-2 virus is highly contagious and spreads from infected chickens to healthy ones through contact and inhalation of their skin dander.

Monilasis, Candidosis, Yeast Infection, or Thrush

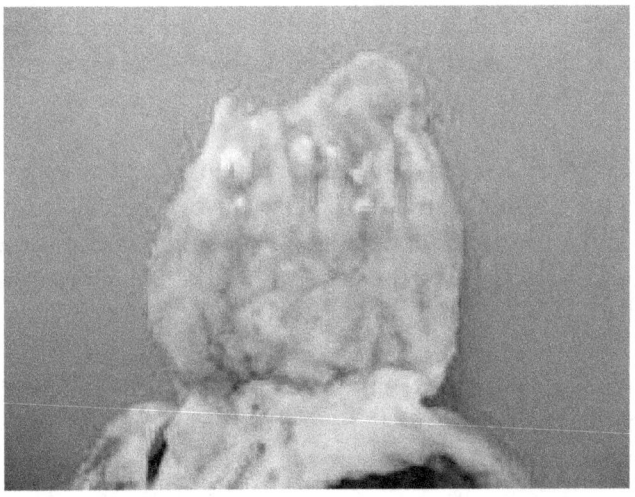

(24)

Type: Fungal "yeast" infection of the mouth and/or cloaca and vent in chickens.

Cause: Fungi, Candida albicans

Source: Contact or ingestion of fungi.

Symptoms: Inflammation, irritation, itching and redness around the mouth or vent, yellowish or white patches of film in the mouth or around the vent area.

Prevention: Treatable with antifungal drugs. Maintaining clean litter, housing, feed and water equipment is the best way to prevent chickens from developing yeast infections.

Candidiasis is an infection caused by the fungi, Candida albicans, also commonly known as a yeast infection. This disease is common in humans but can be present in other animal and bird species.

Symptoms such as inflammation, redness and irritation, soreness, itching, and yellowish or white patches of film may present on the skin and membranes inside the mouth (Thrush), and within and around the vent area in chickens. Yeast infections are typically mild and isolated to one area, and treatable with antifungal drugs. In more severe cases, chickens may experience internal problems like gizzard erosion. Death is uncommon.

Antibiotics should never be used for fungal infections as they can cause resistance buildup against the diseases they can treat and cure. Candidiasis infection should be treated with antifungal drugs, Nystatin, added to feed or water for three to ten days.

Maintaning a clean environment is the best way to avoid infection. By removing old litter and sterilizing feed and water equipment, this fungus can be prevented from reaching the high levels necessary for the onset of an infection.

Mycoplasmas

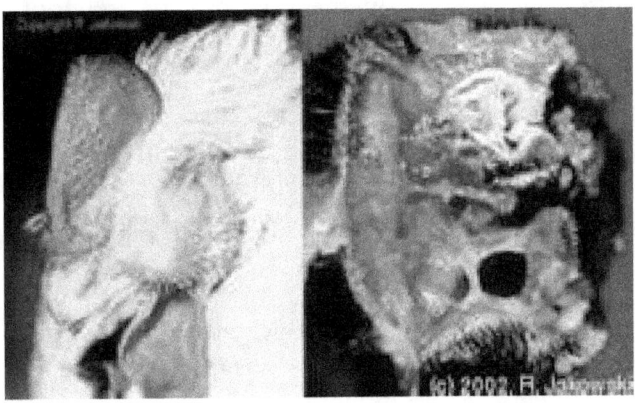

(25)

Type: Respiratory disease caused by bacteria in chickens, worldwide.
Cause: Bacteria, Mycoplasma gallispeticum
Source: Contact with infected birds.
Symptoms: Coughing, nasal discharge, loss of appetite, poor growth and egg production.
Prevention: Tilmicosin, tylosin, spiramycin, tetracycline and fluoroquinolone may be used to treat Mycoplasma infections. Maintaining a clean environment and obtaining chickens through legitimate farmers is the best preventative.

Mycoplasmas are the smallest known cells of bacteria that can replicate independently, with over 100 different species, that can cause infections and diseases in humans, animals, and

57

birds, including chickens. These tiny bacteria are resistant to antibiotics and penicillin because they do not possess cell walls.

Mycoplasma infections in chickens and other birds are caused by the bacterium, Mycoplasma gallisepticum. Infections caused by Mycoplasma gallisepticum cause respiratory disease in chickens and can spread through a flock of healthy birds by close contact. Infected chickens will exhibit coughing, nasal discharge, loss of appetite, poor growth, and poor egg production. Death is rare.

Scientists and poultry health agencies around the world, have worked diligently to eliminate this disease with success. Tilmicosin, tylosin, spiramycin, tetracyclines, and fluoroquinolones may be used to treat Mycoplasma infections.

Obtaining uninfected chicks or chickens through legitimate farmers or traders is critical for prevention. Maintaining a clean and healthy environment is highly effective to prevent infection in backyard flocks.

Newcastle Disease, NDV

(26)

Type: Highly contagious and dangerous bird disease, affecting many species including chickens, worldwide.
Cause: Virus, Avian Paramyxovirus type 1
Source: Ingestion and contact with feces of infected chickens.
Symptoms: Severe nervous and respiratory problems, including twisting of the head and neck. Mortality rate is 90%.

Prevention: Prophylactic vaccinations reduce the chances of an outbreak. Maintaining clean and sanitary living conditions is the most effective preventative.

Newcastle disease is a highly contagious and dangerous bird disease that can affect chickens, and many other domestic and wild bird species, worldwide. The disease is caused by the Avian Paramyxovirus type 1 virus. It was first identified in Newcastle, England in the early 1900s. The Newcastle Disease can spread quickly through an entire flock through ingestion, contact with infected feces, food, water, equipment, and even clothing or shoes of the farmer. The mortality rate for NDV is high with farmers losing 90-100% of their infected flock. Infected chickens should be quarantined as a safety precaution.

Chickens infected with the Newcastle Disease Virus may present with fatigue, coughing, loss of appetite, and reduced egg production that will develop severe respiratory problems and experience trouble breathing, nervous system problems, and death. The symptoms depend on the strain of the virus and the health and age of the infected chicken. Chickens infected with NCV will exhibit symptoms within two to 15 days from the time of infection.
The primary indicator is a rapid deterioration of health and sudden death, including a distorted twisting of the chicken's head and neck. Severe nervous system problems include muscular tremors, drooping or listless wings, circling, swelling of the eyes and neck, and green watery diarrhea.

In acute cases, the death is very sudden. At the beginning of the outbreak, the remaining birds do not seem sick. In flocks with good immunity, however, the signs (respiratory and digestive) are mild and progressive, but on the seventh day nervous symptoms are exhibited like the twisting of the heads.

Necrotic Enteritis

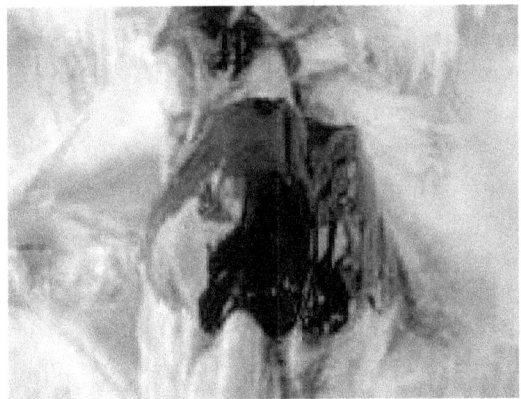

(27)

Type: Bacterial infection of the small intestines in chickens and other bird species, worldwide.
Cause: Bacteria, Clostridium perfringens
Source: Ingestion and contact with contaminated feed and water, and feces.
Symptoms: Listlessness, loss of appetite, closed eyes, ruffled feathers, dark diarrhea, decaying odor, fever, abdominal pain, immobility, and in some cases sudden death.
Prevention: Penicillin can be administered in drinking water, and bacitracin in feed to treat infections in chickens. Disinfecting and maintaining clean litter, housing, feed and water equipment goes a long way in prevention and the spread of this bacteria.

Necrotic Enteritis is a chronic bacterial infection of the small intestines caused by the spore-forming bacterium, Clostridium perfringens. Clostridium perfringens bacterium is commonly associated with food poisoning in humans and birds. When ingested, it attacks the small intestines within 24 hours, producing gas gangrene in the tissues and muscles, accompanied by decomposition and a decaying smell.

Chickens infected with Necrotic Enteritis initially present with fatigue, listlessness, loss of appetite, closed eyes, ruffled feathers, and dark-colored diarrhea. Immobility can occur, and in severe cases even death.

Clostridium perfringens bacteria are common to the environment, and therefore ingestion of non-harmful strains is not uncommon. Serious food poisoning and intestinal infections occur as a result of poorly handled or prepared poultry. Chickens that ingest and develop serious necrotic enteritis infections can transmit harmful bacteria growth to other birds by contaminating feed and water equipment and through their feces.

To treat infected chickens, Penicillin can be administered through drinking water, while bacitracin can be given through feed. These medications are not only used to treat the infection but can also serve as preventative measures to control bacterial growth. Additionally, implementing proper biosecurity practices, such as regular disinfection and maintaining clean litter, housing, and equipment, is crucial to prevent the growth and spread of this bacteria in poultry flocks. By combining effective medication and strict biosecurity measures, the risk of Necrotic Enteritis can be significantly reduced, promoting the overall health and well-being of the flock.

Psittacosis, Parrot Disease, Parrot Fever, Ornothosis, Avian Chlamydiosis, or AC

(Parrot fever) OldVet.com

(28)

Type: Highly contagious and lethal infectious disease in humans, animals and birds, worldwide.

63

Cause: Atypical Bacteria, Chlamydia psittaci
Source: Ingestion, inhalation and/or contact with bacterium in nasal discharge and/or feces of infected birds.
Symptoms: Watery green diarrhea, inflamed eyes, nasal discharge, difficulty breathing and death.
Prevention: Treatment of psittacosis is usually via antibiotics, including: doxycycline or tetracycline administered as water additives, or direct injections.

Psittacosis, also known as avian chlamydiosis or AC, is an infectious bacterial disease that can affect chickens and various bird species worldwide. Infected chickens can remain contagious for several months, during which they release dangerous atypical bacteria in their feces, posing a risk of large outbreaks and significant losses among the flock.

Chlamydia psittaci initially attacks the respiratory system causing nasal discharge and difficulty breathing, before affecting the intestinal tract and other organs. Watery green diarrhea and inflamed eyes are also common indicators. Infected chickens may also exhibit listlessness, fatigue, ruffled feathers, immobility, and even death.

The primary treatment for psittacosis typically involves administering antibiotics such as doxycycline or tetracycline. These antibiotics can be given to chickens through water additives or direct injections, effectively combating the bacterial infection.

To prevent psittacosis outbreaks, it is crucial to obtain chickens from reputable and legitimate farmers who practice proper biosecurity measures. Additionally, maintaining a clean and healthy environment for the flock is vital in reducing the risk of infection and promoting overall bird health.

By implementing these preventive measures and promptly treating any infected birds, chicken owners can safeguard their flocks against the impact of psittacosis and maintain a thriving and disease-free poultry population.

Pullorum Disease, Salmonella Pullorum, "White Diarrhea"

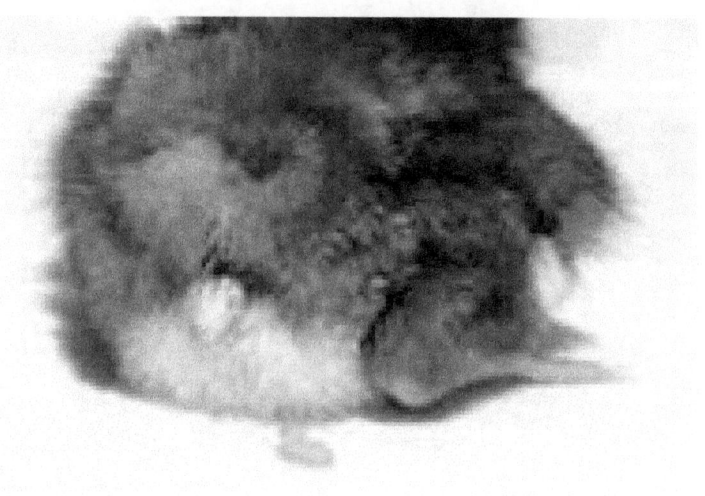

(29)

Type: Bacterial disease in chickens and other bird species, worldwide.

Cause: Bacteria, Salmonella Pullorum

Source: Ingestion and contact with bacteria in cracked or broken eggs, and through contaminated feed and water, and feces.

Symptoms: Listlessness and fatigue, loud chirping, loss of appetite, closed eyes, ruffled feathers, white diarrhea, difficulty breathing, immobility, and in some cases death.

66

Prevention: Vaccines are not normally used for prevention. Amoxycillin, sulponamide, tetracycline, and fluoroquinolone may be used for treatment, and to prevent the spread. Disinfecting and maintaining clean litter, housing, feed and water equipment goes a long way in prevention and the spread of this bacteria. During egg incubation periods, cracked and broken eggs and egg shells should always be disposed of properly to prevent possible infection.

Pullorum Disease is caused by poultry-adapted strains of the bacterium, Salmonella Pullorum. While this disease can affect the health of mature chickens, young chicks up to three weeks of age are more commonly affected and can die of infections setting in during immune development.

The Salmonella Pullorum bacterium is commonly associated with the white part of the egg, or Albumin, and is found growing in cracked and contaminated eggs. During brooding, incubation, and hatchery periods, all broken, cracked, or compromised eggs and their contents should be properly disposed of so that chicks, chickens, other animals, and humans do not ingest this bacterium.

Chickens that ingest these bacteria and develop Pullorum Disease present with white diarrhea as a result of high levels of Salmonella Pullorum bacteria in their feces. They exhibit a wide range of symptoms, including listlessness and fatigue, loud chirping, loss of appetite, closed eyes, ruffled feathers, gasping and difficulty breathing, immobility, and even may lead to death.

Vaccines are not normally used for prevention. Amoxycillin, sulphonamide, tetracycline, and fluoroquinolone may be used for the treatment and prevention of disease. Disinfecting and maintaining clean litter, housing, feed, and water equipment go a long way in preventing and spreading this bacterial disease.

Red Mite, Poultry Mite, Dermanyssus Gallinae

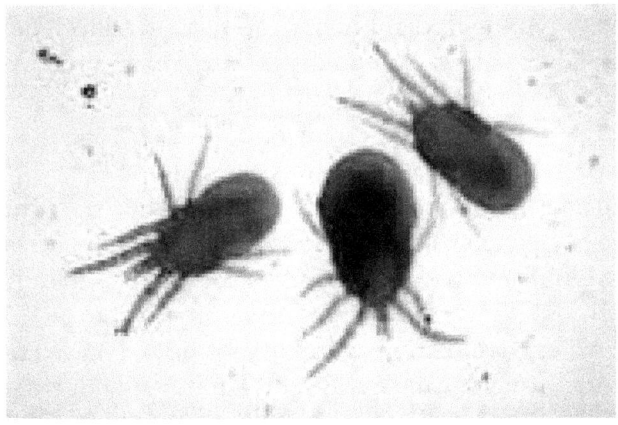

(30)

Type: Red mite infestation of the skin in domesticated birds, including chickens.

Cause: Parasite, Red Mites

Source: Contact with red mites in soil, nesting boxes, cracks and crevices in poultry housing.

Symptoms: Presence of red or grey mites on the skin and within the plumage, irritation, itching and restlessness, skin lesions, anemia (lack of blood), pal combs and wattles, drop in egg production, or eggs produced with spots. Death can occur in chicks during immune development.

Prevention: Multiple strategies are needed to combat Red Mites effectively. Using Ectoparasiticides like Pyrethroids,

69

organophosphates, carbamates, citrus extracts, vegetable oil, mineral-based products, and sand dust is effective for controlling red mites in the environment. Regular inspection of the poultry house is essential to check for signs of Red Mite infestation, such as mites, eggs, and fecal droppings.

Dermanyssus gallinae, also known as red mites or poultry mites, are common parasites in the environment. Mites feed on the blood of their hosts, attacking chickens during periods of rest, predominantly at night. Although they are called red mites, they are most often white or grey and become red as they feed. During daylight hours, red mites tend to hide in the cracks and crevices of the poultry house and lay eggs. As a result, mites can multiply rapidly, causing a large infestation in only a matter of five to seven days.

During an infestation, chickens may experience skin irritation and lesions, itching, and restlessness, and develop illness from lack of blood and bacteria spread by the mites to their hosts. Pale combs and wattles, a drop in egg production, or abnormal egg production, usually visible in the presence of spots on eggshells, can develop if red mites are not treated. Death is uncommon but can occur in young chickens that have not completed the development of their natural immune system.

Red mites can survive up to 10 months in a vacant poultry house. To prevent outbreaks of red mite infestations in chickens, it is crucial to implement multiple measures. Maintaining clean poultry housing is important, and additional strategies such as creosote treatment of wood and fumigation can effectively kill mites hidden in chicken coops. Filling in cracks and crevices is also an essential measure to eliminate hiding places for red mite reproduction.

Scaly Leg, Knemidokoptosis

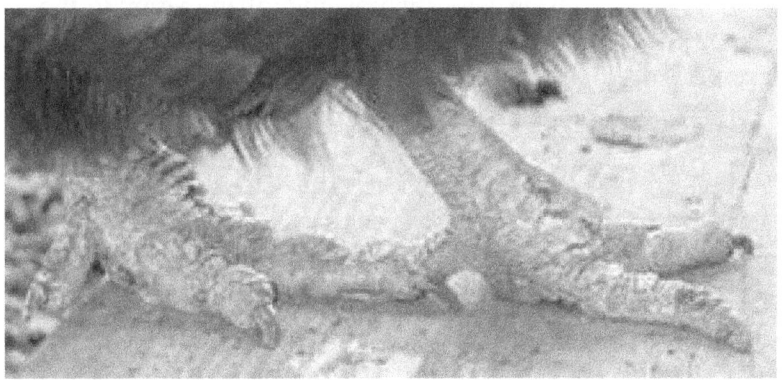

(31)

Type: Parasitic disease caused by mites, common to chickens and other birds resulting in scaly legs.

Cause: Parasite, Knemidocoptes mutans (mites)

Source: Contact with mites in soil, nesting boxes, perches, and poultry housing, and most often as a result of contact with another bird carrying the mites on their bodies.

Symptoms: Raised hard and crusty scales of the legs, irritation, itching, redness and inflammation, white crusty appearance, inflamed wattles and combs, and in the case of extreme infestations, loss of limb usage.

Prevention: Effective treatment and control of scaly leg mites in chickens require appropriate acaricides or veterinary-prescribed treatments specifically designed to target and eliminate the mites. While some home remedies like the use

71

of petroleum jelly, vegetable oil, or chest rubs might create a temporary physical barrier and suffocate some external parasites.

Scaly leg disease is common among chickens and other birds, caused by the parasitic mite, Knemidocoptes mutans. These tiny ticks burrow into the skin of chickens under their scales, leading to inflammation and raised, protruding scales. Knemidocoptes mutans can also infest the wattles and combs of chickens, causing skin irritation and inflammation in those areas. In extreme infestation, it can result in a white crusty appearance on the affected areas of the body, and if left untreated, it can lead to lameness or loss of leg usage.

Mites are commonly found in the soil and other elements of the environment, and some species spend their entire life cycle living in the skin of birds like chickens. Mites thrive in warm and humid conditions and are often associated with poor conditions in poultry housing, such as a lack of proper ventilation.

Effective treatment and control of scaly leg mites in chickens require appropriate acaricides or veterinary-prescribed treatments specifically designed to target and eliminate the mites. While some home remedies like the use of petroleum jelly, vegetable oil, or chest rubs, might create a temporary physical barrier and suffocate some external parasites, they are not the most reliable or comprehensive solutions for eradicating scaly leg mites. Instead, it is essential to use proven acaricides to effectively treat the infestation and prevent further spreading.

In addition to using appropriate treatments, gently brushing infected scales with soapy water can be a helpful method for treating and removing inflamed scales. Regularly cleaning poultry housing to remove loose or dropped scales is also crucial to eliminate potential infestations, as mites can survive up to a month and infect other chickens. Proper insecticide application and thorough cleaning are useful measures to eliminate mites and other ticks from poultry housing.

By combining these strategies, including targeted acaricides, cleaning practices, and proper hygiene, poultry owners can effectively manage and control scaly leg mite infestations, promoting the well-being of their chickens and maintaining a healthier poultry environment.

Toxoplasmosis

(32)

Type: Parasitic disease occurring in warm-blooded animals, including humans and birds, most commonly found within cat feces.

Cause: Parasite, Toxoplasma gondii

Source: Ingestion and contact with feces and expended cysts infected with the internal parasite.

Symptoms: Flu-like symptoms can occur, but most chickens do not present with physical illness. Behavioral changes are the most common indicators, including: listlessness, depression, and brain disorders are the primary indicators, and in rare cases, death.

Prevention: Antibiotics are not effective for the treatment of Toxoplasmosis. The best preventative is maintaining a clean and healthy environment, especially if cats are present in the home and yard.

Toxoplasmosis is a parasitic disease of warm-blooded animals, including humans and birds. The parasite, Toxoplasma gondii is most commonly found in cat feces, and infections are believed to be the result of ingestion of infected fecal material, which transfers the internal parasite, in its second phase, into a new warm-blooded host.

Toxoplasmosis in chickens results when this parasite invades internal cells, reproduces, and infects tissues of the muscles and brain, forming cysts. Flu-like symptoms can occur, but most chickens do not present with any physical illness. Toxoplasmosis most commonly affects the brain and causes behavioral changes, including listlessness, disorientation and depression, fear, and brain disorders. In rare cases, death can occur but is usually a result of a compromised immune system, commonly associated with other diseases or illnesses already present in the chicken.

Trichomoniasis, Canker, Frounce

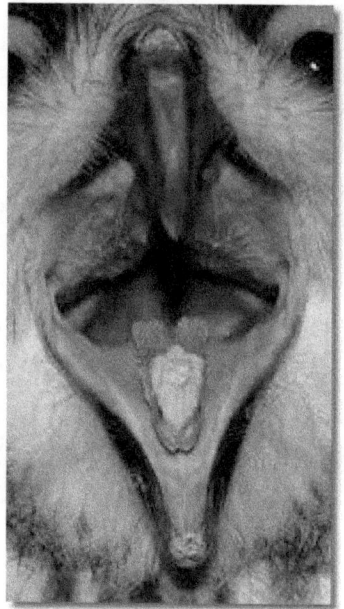

This picture shows an Eurasian Eagle Owl
with the characteristic lesions on the
tongue, before treatment.

© Ian Berwick   Used by permission

(33)

Type: Parasitic disease occurring in young birds, including chicks.
Cause: Parasite, Trichomonae gallinea
Source: Ingestion and contact with infected feces.

Symptoms: Mouth open, drooling, continuous swallowing, cheese-like plaque deposits in the mouth and around the beak, loss of appetite, watery eyes, and even death.

Prevention: 2-amino-5-nitrothiazole is widely used to treat this disease with no known resistance. Introduction of new birds from reputable breeder or farmer, and removal of all stagnate water sources in outdoor run areas are best preventatives.

Trichomoniasis is a disease caused by the parasite, Trichomonas gallinae, that primarily affects young birds, including chickens. This disease can also be referred to as canker in pigeons and Frounce in falcons and other birds of prey.

This protozoan parasite is commonly found in stagnant water, but its transmission to chickens is not typically through water sources. Instead, it is primarily transmitted through direct contact with infected birds or contaminated food. Once inside a young bird host, the parasite multiplies quickly, infecting the nasal cavity, mouth, and respiratory tract. Digestive tracts may also be affected, but unlike other parasites, Trichomonas gallinae die as they are passed outside of the host, through the feces.

Young chickens or chicks infected with Trichomoniasis present with open mouths, drooling, continuous swallowing, and gasping. They often exhibit cheese-like plaque deposits in the mouth and around their beaks, have watery eyes, and appear tired, listless, and may stop feeding. Death can occur in young chicks, particularly those with underdeveloped immune systems.

Treatment option for Trichomoniasis is 2-amino-5-nitrothiazole, which is widely used with no known resistance and can show positive results within one to two days.

The best preventative measure against Trichomoniasis is maintaining a clean and healthy environment, free of any stagnant water sources, in or around the yard where these parasites can grow. Bird baths, especially those shared by different bird species, should be dried out, disinfected, and left dry for several weeks to ensure the removal of parasites after infections. Additionally, practicing good hygiene and preventing direct contact with infected birds can help reduce the risk of transmission.

Ulcerative Enteritis

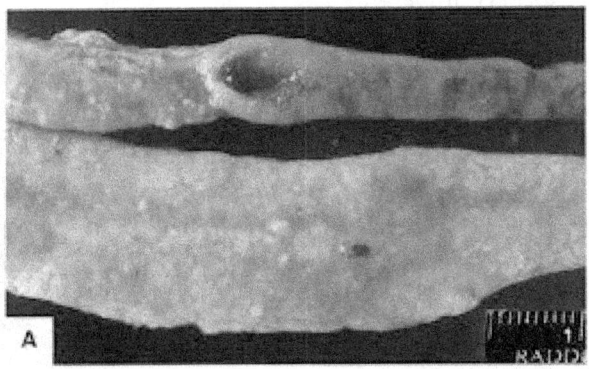

Ulcerative Enteritis   OldVeT.com

(34)

Type: Bacterial infection of the intestines in birds, including chickens, worldwide.

Cause: Bacteria, Clostridium colinum

Source: Ingestion and contact with infected feces, or carrier birds.

Symptoms: Weight loss, loss of appetite, ruffled feathers and hump backs are primary indicators. As the infection advances, ulcers and lesions in the intestines cause chicken droppings to appear white in color, or result in watery diarrhea.

Prevention: Feed additives like Streptomycin and furazolidone are effective in treating Ulcerative Enteritis in chickens. Overcrowding and the introducing of new birds to

79

a flock are common factors in outbreaks. To prevent and manage the disease, it is essential to maintain healthy living conditions for existing flocks and acquire new chickens from reputable dealers and farmers.

Ulcerative Enteritis is a bacterial infection of the intestines in birds, including chickens, caused by the spore-forming bacterium Clostridium colinum. The disease is highly contagious and can spread to healthy birds through carriers and ingestion of infected feces.

Chickens infected with Clostridium colinum may present with symptoms such as depression, reduced activity, increased thirst, and sometimes blood-stained droppings. Ulcers and lesions within the intestinal tract are characteristic of the disease, but other symptoms mentioned, such as weight loss, ruffled feathers, humpbacks, and white droppings, are more commonly associated with other poultry diseases like Coccidiosis.

Yolk Sac Infection, Omphallitis

(35)

Type: Bacterial infection occurring in the navel of newly hatched chicks, worldwide.
Cause: Bacterium, E. coli, Staphylococci, Proteus, Pseudomonas.
Source: Bacterial contamination of developing eggs during incubation, most often due to unsanitary conditions in nesting boxes.
Symptoms: Swollen abdomen, water retention, loss of appetite and slow growth, and slow healing navels, including

tags of yolk hanging from the navel on newly hatched chicks. Mortality is high in affected chicks.

Prevention: Maintaining clean and sanitary nesting and chick boxes is highly effective in preventing bacterial infections of eggs during the incubation and hatching stages.

Yolk Sac Infection or Omphallitis is a bacterial infection in newly hatched chicks, affecting bird species, including chickens, worldwide. This disease occurs as a result of bacterial contamination during the first couple of days of incubation, disrupting the healing process of navels on newly hatched chicks.

Omphalitis is caused by several bacteria strains, including E coli, Staphylococci, Proteus, and Pseudomonas, and is commonly associated with poor hatchery hygiene. Unsanitary conditions in nesting boxes and poultry houses, during brooding, incubation, and hatchery result in unhealthy offspring, and often large losses in new chicks that haven't had proper time to develop strong immune systems.

Chicks born with Yolk Sac Infection do not grow, feed or develop properly. They often do not open their eyes, and present with swollen abdomens, and tags or flaps of yolk hanging from their unhealed navel. Chicks with bacterial infections of the navel are not able to properly heal or develop. Antibiotics can be used for treatment, but infected chicks will typically die within the first 7 days of their life.

Prevention revolves around hygiene. Maintaining good clean nesting boxes through egg laying and incubation, and chick boxes after hatching is the key to the development of healthy chicken offspring. Frequent litter changes, disposal of heavily soiled, broken, or floor eggs, and disinfecting and maintaining a healthy sanitary environment for hens and new chicks is highly effective as a preventative.

Chapter 3 – Maintaining Good Chicken Health

Maintaining good chicken health, and preventing the occurrence of infection, illness or disease can be achieved by following basic principles in the acquisition, housing, and care you provide. Whether you are a small back-yard farmer of a few chickens for egg or meat production, or manage a poultry house for commercial purpose, the basic guidelines for prevention are as follows:

Healthy Stock

Many of the diseases common to chickens are also common to other bird species, animals, including livestock and pets, and humans. Many of these diseases have mutated and adapted from one species to another, have many different strains, spread easily, and require different treatments and preventative methods. As such, many health agencies around the world have studied and learned how to effectively treat and/or eliminate many of the more dangerous elements that can affect the health of your chicken(s), and/or flock. Some of these common, but highly infectious and dangerous health issues have been completely eliminated in some regions around the world.

It is critical that chickens are obtained through legitimate and reputable breeders and farmers to completely avoid many of the health problems detailed in this book. Many birds, including chickens can remain carriers of bacteria, fungi, parasites, and viruses, after they have recovered from illness, and spread infection, illness and disease to healthy flocks, completely undetected. Reputable farmers and dealers must follow guidelines mandated by national and international health agencies to eliminate the spread of destructive and deadly health problems in chickens and other animals throughout the world. By obtaining eggs, chicks or chickens from health-conscious farmers, many of these health problems can be avoided entirely. It is even more critical to be diligent about the source of new chickens, when introducing them into an existing flock, as many infections, illness and disease spread undetected and quickly from chicken to chicken, and may not only affect the health, but lives of the entire flock.

Proper and Adequate Chicken Housing

Inadequate and unsanitary chicken housing is the source and primary culprit for the growth and spread of many dangerous and toxic elements to chickens and humans alike. Building or buying, and maintaining proper poultry housing is critical to maintaining good overall health in your flock, and preventing an outbreak of illness or disease.

Proper and adequate chicken housing should include the following:

1.   Shelter from the Elements
Proper poultry housing should provide birds with physical protection from extreme cold weather, extreme heat, humidity, sunlight, rain and snow.

2.   Good Ventilation

Poultry housing, regardless of structure type, must be well-ventilated. The circulation of fresh outside air within the chicken coop is critical for the absorption of excess humidity and toxic fumes from chicken excrement. Good ventilation is also important during hot summer months for the elimination of hot stale air that can cause heat stress and stroke. Good ventilation can keep coop temperatures cooler than outside temperatures by 10-degrees. As a rule of thumb, if humans are unable to breath within the poultry house, the chickens are too.

3.   Absorption and Removal

Straw, wood chips and shavings, and other natural porous materials are commonly used for chicken coop floor covering and nesting boxes. These natural litters are inexpensive, easy to find and replace, and exceptionally good for the absorption and removal of chicken droppings, and the dangerous bacteria, fungi, parasites and viruses that can grow within them inside chicken houses. The use of natural litters is also exceptionally useful for trapping and controlling excess moisture and humidity from within the coop. Once natural litters are saturated, they not only lose their effectiveness, but become a breeding ground for disease. Old litter should be removed and replaced on a regular basis, at least once a month.

4.   Adequate Space

Adequate poultry housing means adequate space for chickens to sleep, and to lay eggs. Adequate spacing is even more critical for brooding during incubation, hatching new chicks, and the growth and development of babies within the coop. As most hens and roosters prefer to perch when they sleep, providing an adequate number of perches for each chicken in the flock is a good rule of thumb to follow. One nesting box should be provided for every four laying hens. Separate chick boxes should be provided during hatchery. Indoor and outdoor runs must be provided for chickens to live healthy. Most outdoor runs simply consist of a fenced-in yard or enclosed area where chickens can naturally forage for food in their environment, and often do not require any maintenance, as excrement is absorbed into the soil as a natural fertilizer.

5.   Regular Cleaning and Maintenance

Chicken coops require minimal cleaning and maintenance. Cleaning is often associated with litter replacement to maintain proper absorption and promote good ventilation within the coop. Cleaning coops once a month thoroughly to remove any bacteria, fungi, parasites growing within the environment is important to ensuring good health in chickens in the long-term. If illness, infection or disease occurs within a chicken flock, or with other animals living in close proximity, litter should be replaced completely as a preventative measure to stop the spread to other chickens. Disinfecting floors, nesting boxes, perches, food and water equipment, is recommended at the first sign of illness, and once every three months as a preventative to keep flocks healthy.

Proper Chicken Care

Chickens are low-maintenance, and require very little effort for on-going care. Proper care requires following basic principles, which are also effective in preventing the outbreak of disease or illness in a flock.

The following principles should be followed to maintain good health in the day-to-day care of chickens:

1.   Feed and Feeder

Proper feed should be provided as a supplement to the chicken's natural diet of insects, weeds, grass, minerals in soil, and small lizards. Poultry feed is divided into four primary categories, depending on age and use of the chicken.
- Complete
- Starter

- Finisher
- Developer

"Complete feed" contains all of the protein, energy, vitamins, minerals and other nutrients necessary for chickens to grow at a healthy and adequate rate, become good consistent egg producers, and live a long and healthy life, free of disease. The primary ingredients in complete feed include: protein, calcium and phosphate. Chickens need more or less of these three ingredients during their development cycle.

"Starter feed" contains all of the necessary vitamins and minerals to promote healthy growth and immune system development in baby chickens, or chicks. Starter feed should be fed to new chicks from time of hatching, until they are six to eight weeks of age. Starter feed contains a higher level of protein and energy than complete feed, important during early development in birds.

A "finisher feed or diet" is provided to young broilers, chickens raised for their meat, from six or eight weeks of age, until they are at slaughter weight. Finisher diets are designed to promote faster growth of muscle mass.

A "developer feed or diet" is fed to pullets or cockerels, chickens less than a year of age, for the first year of their development. The development feed is designed to promote healthy reproductive systems for good steady egg production, and ensure strong egg shells and quality in laying hens.
 As a rule of thumb, fresh "complete feed" should be supplied based on number of chickens in flock, during daylight hours, in outside runs each day. Feeders can be filled first thing in the morning so chickens can forage and feed throughout the day. As dusk sets in, excess feed should be removed to prevent spoilage, food poisoning and fungi or bacterial growth, and health problems as a result.

Feeders should be cleaned on a regular basis, and disinfected to kill any bacteria, fungi, or parasites that may have contaminated the feeding device naturally.

2.   **Water and Watery**

Chickens each need a good supply of fresh water on a daily basis. Keeping chickens well hydrated can go a long way in the prevention of illness. Watery or watering device should be cleaned and disinfected on a regular basis, along with the feeder, to kill any bacteria, fungi, or parasites that may have contaminated the watery device naturally. A watery device that shows signs of rust should be eliminated completely.

If symptoms or signs of illness present in a chicken, the following steps can help eliminate the spread of disease throughout a flock, and speed up chicken recovery:

1.   **Separate Sick from Healthy**

If a chicken shows sign of sickness, it is best to separate that chicken from healthy birds for a few days, at least. Symptoms of infection, illness and disease may take several days to fully present. Separating sick from healthy birds can help contain outbreaks of disease.

2.   **Medicines**

There are many antibiotics, food and water additives that can prevent and/or treat illness and disease in chickens. Over-use or incorrect use of antibiotics and medicines can build up resistance in chickens, against different strains of illness. Always carefully research and follow administration of any drug to chickens, carefully and diligently. Chickens often recover from illness on their own, through vitamin and mineral supplements to their water or feed, and by the support of a thorough cleaning and disinfecting of their environment, fresh water, and rest.

3.   **Proper Disposal**

In the event of an outbreak of infection, illness or disease, proper disposal of dead chickens is critical, not only for the safety of remaining members of an existing flock, but for animals of all species that live in and around your environment, including humans. Chickens that die from disease should always be burned. Cracked and broken eggs should also be properly disposed of so that other bird and animals species do not ingest them and fall sick. While chickens will scavenge on human compost of naturally discarded food waste, care should be taken to prevent ingestion of rotting or potentially dangerous bacteria in animal products, including eggs and meat.

Poultry Medications

There are numerous medications and products on the market to prevent and treat common chicken infection, illness and disease.

Poultry Disinfectants

There are numerous disinfectants that work effectively in sterilizing your indoor and outdoor chicken runs to prevent the growth of fungi, bacteria and the spread of dangerous virus. Poultry disinfectants come in a variety of forms, including aerosol sprays, defoggers, and concentrated liquids. Poultry disinfectants can be applied directly to the chicken coop, feeding equipment, watery systems, and outdoor runs to eliminate the tiny disease-causing living organisms. Chickens should be relocated to an outdoor run or area away from the chemicals during fumigation and heavy cleaning. Disinfectants should be used during regular cleaning and maintenance of chicken coop, equipment and runs, recommended at least once every one to three months.

External Parasite Prevention

There are a variety of products available for the treatment of lice, fleas, ticks, mites, and other parasites common to the environment, that can negatively affect the health of your chicken(s). When chickens are able to live free of parasites that burrow into their skin and scales, and feed on their blood, like fleas, mites, ticks and lice, they have a happier and healthier disposition, which leads to natural weight gain and a good steady egg production. There are numerous brands of sprays and concentrates on the market that can be applied directly to the exterior of the chicken, their bedding, housing and even outdoor runs. In addition, fossil shell flour or diatomaceous earth is a dust, highly effective as a food additive that not only kills unwanted parasites, but also improves egg shell quality and production. Fossil shell flour can also be provided in a small shallow tray or pan to allow chickens to dust bathe naturally, eliminating fleas and lice from their plumage, as well.

Poultry Worm Prevention

In addition to sprays and dust, chickens can be given a wormer medication to prevent internal parasitic infections in the form of worms. There are numerous products, including tablets administered to each chicken individually once a month, or dissolved in their water.

Poultry Vaccines

There are only a few vaccines available for a few of the common diseases chickens can encounter. Vaccines are typically recommended for chickens used for egg production, and not for broilers or chickens produced for their meat. Vaccines are currently available for New Castle Disease and Fowl Pox, and allow small back-yard farmers the option of injecting or administering the vaccines at home, versus the timely and costly process of visiting a veterinarian or animal doctor. New Castle Disease vaccines are typically administered to new baby chicks between 2 and 15 weeks of age.

Poultry Blood Tests

Do-it-Yourself blood test kits are also available for the detection of Salmonella Pullorum and Fowl Typhoid, two dangerous and often deadly bacterial diseases. These blood test kits allow small farmers to test new chickens for disease, prior to introducing them into an existing flock, to prevent a devastating outbreak.

Poultry Antibiotics

There are numerous antibiotics available for the treatment of chicken disease. Most antibiotics are provided in the form of a feed or water additive.

Poultry Vitamins and Supplements

Poultry vitamins and mineral supplements are available in the form of feed and water additives and can be administered to chickens on a regular basis to promote growth and maintain good health. However, many small farmers use vitamin electrolytes and minerals during breeding season, extreme weather, and if chickens begin exhibiting depression, listlessness, or signs of the onset of illness, infection or disease. Supplementing chickens with natural vitamins and minerals can be highly effective in fighting of disease.

Nutritional deficiencies can present in a wide array of symptoms and health problems in chickens. Some of the most common uses for specific vitamin supplements include:

•   Rickets - Deficiency of vitamin D3, calcium and/or phosphorus. Add cod liver oil and DiCal, or steamed bone meal to diet.
•   Crazy chick disease – Deficiency of vitamin E. Add source of pure vitamin E to diet.

- Curly Toe Paralysis - Deficiency of riboflavin. Add milk products to diet.
- Perosis or slipped tendon - Deficiency of choline, manganese, and/or biotin. Add choline, manganese, and/or biotin to diet.
- Pale birds – Deficiency of vitamin A. Add cod liver oil.

Most Common Poultry Feed and Water Additive Medications

In addition to the specific medications listed under each disease summary in the previous chapter, the following medicines are most commonly used by farmers all over the world, to prevent and treat common health problems in chickens, as follows:

- **Terramycin** (Oxytetracycline) – water additive.
- **Aureomycin** (Chlortetracycline) – water additive.
- **NF 180** – feed additive.
- **Neomycin** – water or feed additive.
- **Gallimycin (**Erythromycin) - water or feed additive.
- **Amprolium** (Corid)
- **Sulfaquinoxaline** or **Sulfamethazine** - water or feed additive.
- **Tramizol** – water additive.

Egg Withdrawal During Medication

Different medicines require different lengths of time for use, and also different lengths of time for the safe elimination of that medicine from the chicken's system and body. The withdrawal time is an important factor to consider, especially as it pertains to hens used for egg production, when those eggs are consumed by humans, or other animals. Eggs produced during the administration of internal evasive medications like antibiotics, penicillin, food and water additives and injections, should be removed and disposed of.

Veterinarians will typically advise of the specific number of withdrawal days, from 1 to 28. Eggs produced during periods of heavy medication may not be safe for consumption and should be disposed of as a precaution to avoid any side effects or health issues in humans.

Are your feathered friends getting the ultimate coop experience they deserve?

We know how much you love your chickens, and we're excited to reveal something that will elevate their coop living to a whole new level! Introducing our meticulously designed coop plans that will not only enhance your chickens' comfort but also make your life as a chicken keeper a breeze!

Whether you're a seasoned poultry enthusiast or just starting your backyard flock, these coop plans are tailored to suit every need and budget. From spacious layouts to smart ventilation systems, predator-proof designs to nesting box perfection – we've got it all covered!

Don't let your chickens miss out on the chance to live in a coop that's truly the envy of the henhouse community. Check the link below to explore our coop plans and discover a world of happy clucks and contented coops.

https://buildingchickencoopsguide. com/

References

The references and links at the bottom of each chicken disease or illness summary, and listed in this section were not only used in the creation of this book, but have been included to provide readers with access to a large quantity of additional resources on chicken diseases, including photos.

The Poultry Site: Diseases of Poultry, by Ivan Dinev, DVM, PhD
http://www.thepoultrysite.com/publications/6/diseases-of-poultry

Wikipedia, The Free Encyclopedia: Chicken – Chicken Disease Table & Links
http://en.wikipedia.org/wiki/Chicken

MSU Cares.Com: Poultry Feed and Nutrition
http://msucares.com/poultry/feeds/poultry_feeds.html

Aspergillosis Reference:
http://en.wikipedia.org/wiki/Aspergillosis
http://www.thepoultrysite.com/publications/6/diseases-of-poultry/212/aspergillosis
http://oldvet.com/wp-content/uploads/2011/05/Asperagellosis.jpg

Bird Flu Reference:
http://en.wikipedia.org/wiki/Avian_influenza
http://partnersah.vet.cornell.edu/avian-atlas/sites/agilestaging.library.cornell.edu.avian-atlas/files/avian_atlas_assets/3.5.08.DSC00210%20x750.jpg

Blackhead Disease References:
http://en.wikipedia.org/wiki/Blackhead_disease

http://www.thepoultrysite.com/publications/6/diseases-of-poultry/207/histomonosis
http://www.ecologyandsociety.org/vol9/iss1/art5/figure1.jpg

Botulism References:
http://en.wikipedia.org/wiki/Botulism
http://www.thepoultrysite.com/publications/6/diseases-of-poultry/187/botulism
http://www.thepoultrysite.com/publications/images/image_Page_021_Image_0013.jpg

Bumblefoot or Ulcerative Pododermatitis Reference:
http://en.wikipedia.org/wiki/Ulcerative_pododermatitis
http://adlib.everysite.co.uk/resources/000/012/843/poultry_litter_fig2b.jpg

Campylobacteriosis Reference:
http://en.wikipedia.org/wiki/Campylobacteriosis
http://www.health-pic.com/EX/09-20-01/Campylobacteriosis.jpg

Coccidosis Reference:
http://img703.imageshack.us/img703/6914/13022011142.jpg

Erysipelas Reference:
http://en.wikipedia.org/wiki/Erysipelas
http://www.backyardchickens.com/image/id/5728486

Fowl Cholera Reference:
http://www.michigan.gov/dnr/0,1607,7-153-10370_12150_12220-26650--,00.html
http://vethomopath.com/fowl.jpg

Fowl Pox References:
http://en.wikipedia.org/wiki/Fowlpox

http://en.wikipedia.org/wiki/Avipoxvirus

Fowl Typhoid References:
http://www.worldpoultry.net/diseases/fowl-typhoid-d98.html
http://www.thepoultrysite.com/diseaseinfo/130/salmonella-gallinarum-fowl-typhoid
http://ts1.mm.bing.net/images/thumbnail.aspx?q=47099949
69760668&id=5bd1f901d5ee4d96cd2769ced7af15d2

Gallid or avian herpesvirus 1 References:
http://en.wikipedia.org/wiki/Gallid_herpesvirus_1
http://ts1.mm.bing.net/images/thumbnail.aspx?q=45060871
19913172&id=34587d7eef279381439fe9de8fd7f813

Gapeworm or Syngamus Trachea Reference:
http://en.wikipedia.org/wiki/Gapeworm
http://www.agriculture.gov.ie/media/migration/animalhealt
hwelfare/labservice/cvrlimages/August%20Figure%203%20
with%20Arrow-400x300.JPG

Infectious Bronchitis Reference (includes extensive vaccine listing):
http://edis.ifas.ufl.edu/ps039
http://en.wikipedia.org/wiki/Coronavirus
http://www.poultrymed.com/Poultry/UploadFiles/PGallery
/3589673850_Big.jpg

IBD References:
http://www.gumboro.com/disease/
http://en.wikipedia.org/wiki/Infectious_bursal_disease
http://www.chickclinicegypt.com/gum.JPG

Infectious Coryza, IC, Coryza, Roup References:

http://www.lah.de/typo3temp/pics/c630af5de1.jpg

Lymphoid Leukosis Reference:
http://en.wikipedia.org/wiki/Avian_leukosis_virus
http://www.thepoultrysite.com/publications/images/image_
Page_057_Image_0001.jpg

Marek's Disease Reference:
http://en.wikipedia.org/wiki/Marek's_disease
http://www.poultrydisease.ir/Atlases/avian-
atlas/sites/agilestaging.library.cornell.edu.avian-
atlas/files/avian_atlas_assets/PAST-042A%20x420.jpg

Monilasis Reference:
http://www.thepoultrysite.com/diseaseinfo/27/candidiasis-
moniliasis-thrush
http://en.wikipedia.org/wiki/Moniliasis
http://www.thepoultrysite.com/publications/images/image_
Page_075_Image_0005.jpg

Mycoplasmas References:
http://en.wikipedia.org/wiki/Mycoplasmas
http://www.thepoultrysite.com/diseaseinfo/94/mycoplasma
-gallisepticum-infection-mg-chronic-respiratory-disease-
chickens
http://ts1.mm.bing.net/images/thumbnail.aspx?q=50082031
09565668&id=436eeff3fe867fb30c1de7d4e5d9f293

Newcastle Disease Reference:
http://en.wikipedia.org/wiki/Newcastle_disease
http://ts4.mm.bing.net/images/thumbnail.aspx?q=48738307
55271599&id=d5a2b1115c388cac529882b4ed37364f

Necrotic Enteritis Reference:
http://www.thepoultrysite.com/diseaseinfo/101/necrotic-
enteritis
http://en.wikipedia.org/wiki/Clostridium_perfringens

http://ts3.mm.bing.net/images/thumbnail.aspx?q=48862690
02646730&id=460ade6003e4f0f44cde26b210807f93

Psittacosis Reference:
http://en.wikipedia.org/wiki/Psittacosis
http://en.wikipedia.org/wiki/Chlamydophila_psittaci
http://ts2.mm.bing.net/images/thumbnail.aspx?q=49470985
86842289&id=9e0640fe1f64d2bf7d647e690e614bf3

Salmonella Reference:
http://www.thepoultrysite.com/diseaseinfo/131/salmonella-
pullorum-pullorum-disease-bacillary-white-diarrhoea
http://en.wikipedia.org/wiki/Salmonella
http://ts3.mm.bing.net/images/thumbnail.aspx?q=50323494
11640586&id=7fac30cdf1592b0d24f2015c9668b339

Red Mite Reference:
http://www.thepoultrysite.com/diseaseinfo/120/red-mite-
and-northern-fowl-mite
http://en.wikipedia.org/wiki/Dermanyssus_gallinae
http://ts3.mm.bing.net/images/thumbnail.aspx?q=47171761
49707194&id=c348f33aa81bbfb3397b4dbcd905c2d6

Scaly Leg Reference:
http://en.wikipedia.org/wiki/Scaly_leg
http://ts2.mm.bing.net/images/thumbnail.aspx?q=45933050
00281513&id=216adfeedfed3348f2d74c4ba907e67a

Toxoplasmosis Reference:
http://en.wikipedia.org/wiki/Toxoplasma_gondii
https://encrypted-
tbn2.google.com/images?q=tbn:ANd9GcQksOt6fqwNhF-
cnnlsCcXRiAG5Hk_DUffahL5Jz5S3WHD6q7q7

Trichomoniasis Reference:
http://www.thepoultrysite.com/diseaseinfo/154/trichomoni
asis-canker-frounce

http://en.wikipedia.org/wiki/Trichomonas_gallinae
https://encrypted-
tbn0.google.com/images?q=tbn:ANd9GcTyZ0aSEURknfJrr
wBUSchEebNSHkU2Vq_cAAdrTGC1yHBXQwTAXw

Ulcerative Enteritis Reference:
http://www.merckvetmanual.com/mvm/index.jsp?cfile=ht
m/bc/201500.htm
https://encrypted-
tbn3.google.com/images?q=tbn:ANd9GcRLbe0g98VJIu4_fl
HYW1es-yFYIBO5LJqV8LDfZySyoysDAXAb

Yolk Sac Infection Reference:
http://www.thepoultrysite.com/diseaseinfo/169/yolk-sac-
infection-omphallitis
http://www.thepoultrysite.com/publications/images/image_
Page_005_Image_0006.jpg

(1)http://images.suite101.com/2911403_COM_salmonella_bacteria.jpg

(2) http://www.poultryhub.org/wp-content/uploads/2012/04/450px-Campylobacter.jpg

(3) http://www.poultrymatters.com/photopost/data/500/medium/fungus_close_up.jpg

(4)http://images.suite101.com/1738877_com_coccidia.jpg

(5) http://www.sxc.hu/photo/1295739

(6) http://oldvet.com/wp-content/uploads/2011/05/Asperagellosis.jpg

(7) http://partnersah.vet.cornell.edu/avian-atlas/sites/agilestaging.library.cornell.edu.avian-
atlas/files/avian_atlas_assets/3.5.08.DSC00210%20x750.jpg

(8) http://www.ecologyandsociety.org/vol9/iss1/art5/figure1.jpg

(9) http://www.thepoultrysite.com/publications/images/image_Page_021_Image_0013.jpg

(10) http://adlib.everysite.co.uk/resources/000/012/843/poultry_litter_fig2b.jpg

(11) http://www.health-pic.com/EX/09-20-01/Campylobacteriosis.jpg

(12) http://img703.imageshack.us/img703/6914/13022011142.jpg

(13) http://www.backyardchickens.com/image/id/5728486

(14) http://vethomopath.com/fowl.jpg

(15)
http://ts1.mm.bing.net/images/thumbnail.aspx?q=4709994969760668&id=5bd1f901d5ee4d96cd2769ced7af15d2

(16)
http://ts2.mm.bing.net/images/thumbnail.aspx?q=5032349411640581&id=c7360a7ebbd57651df2e40fee5685e30

(17)
http://ts1.mm.bing.net/images/thumbnail.aspx?q=4506087119913172&id=34587d7eef279381439fe9de8fd7f813

(18)
http://www.agriculture.gov.ie/media/migration/animalhealthwelfare/labservice/cvrlimages/August%20Figure%203%20with%20Arrow-400x300.JPG

(19) http://www.poultrymed.com/Poultry/UploadFiles/PGallery/3589673850_Big.jpg

(20) http://www.chickclinicegypt.com/gum.JPG

(21) http://www.lah.de/typo3temp/pics/c630af5de1.jpg

(22)
http://www.thepoultrysite.com/publications/images/image_Page_057_Image_0001.jpg

(23) http://www.poultrydisease.ir/Atlases/avian-atlas/sites/agilestaging.library.cornell.edu.avian-atlas/files/avian_atlas_assets/PAST-042A%20x420.jpg

(24)
http://www.thepoultrysite.com/publications/images/image_Page_075_Image_0005.jpg

(25)
http://ts1.mm.bing.net/images/thumbnail.aspx?q=5008203109565668&id=436eeff3fe867fb30c1de7d4e5d9f293

101

(26)
http://ts4.mm.bing.net/images/thumbnail.aspx?q=4873830755271599&id=d5a2b1115c388
cac529882b4ed37364f

(27)http://ts3.mm.bing.net/images/thumbnail.aspx?q=4886269002646730&id=460ade6003
e4f0f44cde26b210807f93

(28)
http://ts2.mm.bing.net/images/thumbnail.aspx?q=4947098586842289&id=9e0640fe1f64d2
bf7d647e690e614bf3

(29)
http://ts3.mm.bing.net/images/thumbnail.aspx?q=5032349411640586&id=7fac30cdf1592b
0d24f2015c9668b339

(30)
http://ts3.mm.bing.net/images/thumbnail.aspx?q=4717176149707194&id=c348f33aa81bbf
b3397b4dbcd905c2d6

(31)http://ts2.mm.bing.net/images/thumbnail.aspx?q=4593305000281513&id=216adfeedfe
d3348f2d74c4ba907e67a

(32) https://encrypted-tbn2.google.com/images?q=tbn:ANd9GcQksOt6fqwNhF-
cnnlsCcXRiAG5Hk_DUffahL5Jz5S3WHD6q7q7

(33) https://encrypted-
tbn0.google.com/images?q=tbn:ANd9GcTyZ0aSEURknfJrrwBUSchEebNSHkU2Vq_cAA
drTGC1yHBXQwTAXw

(34) https://encrypted-
tbn3.google.com/images?q=tbn:ANd9GcRLbe0g98VJIu4_flHYW1es-
yFYIBO5LJqV8LDfZySyoysDAXAb

(35)
http://www.thepoultrysite.com/publications/images/image_Page_005_Image_0006.jpg

www.ingramcontent.com/pod-product-compliance
Lightning Source LLC
Chambersburg PA
CBHW071722170526
45165CB00005B/2114